 Advances in World Aquaculture, Volume 1
Managing Editor, Paul A. Sandifer

Farming Bivalve Molluscs: Methods for Study and Development

by

D. B. Quayle

Department of Fisheries and Oceans
Fisheries Research Branch
Pacific Biological Station
Nanaimo, British Columbia V9R 5K6
Canada

and

G. F. Newkirk

Department of Biology
Dalhousie University
Halifax, Nova Scotia B3H 4J1
Canada

Published by

THE WORLD AQUACULTURE SOCIETY
in association with
THE INTERNATIONAL DEVELOPMENT RESEARCH CENTRE

The World Aquaculture Society
16 East Fraternity Lane
Louisiana State University
Baton Rouge, LA 70803

Library of Congress Catalog Number: 89-40570

ISBN: 0-9624529-0-4

ACKNOWLEDGMENTS

The following figures are reproduced with permission: Figures 1 – 10, 12, 13, 17, 20, 22, 23, 32, 35, 37, 42, 45, 48, 50 – 54, 62, 64, 72, 75, 86, and 87 from the Fisheries Board of Canada; Figures 11 and 21 from the United States Government Printing Office; Figure 15 from the Buckland Foundation; Figures 18, 19, 24 – 28, 33, 34, 38, 41, 56, and 65 from the International Development Research Centre; Figures 29 and 30 from the Journal of Shellfish Research; and Figure 43 from Fritz (1982). Data for Figure 39 are used with permission from D. Thompson, for Figure 40 from W. M. Indrasena, and for Figure 44 from A. G. Elnaiem. Figures 57 and 58 are based on a Canadian Hydrographic Survey chart and Figures 59 and 60 on British Admialty charts and are reproduced with permission.

PREFACE

Bivalve mollusc farming has been developing rapidly in many temperate countries in recent years. This development is, to a large extent, driven by market demand. Interest in the development of tropical bivalve species is also high where there is potential for bivalves to provide cash crops and additional animal protein. However, bivalve farming is constrained in many countries by a limitation of available biological information and experience in developing and testing culture systems. It is for this reason that the International Development Research Centre (IDRC) (Canada) has been sponsoring research projects in developing countries of Asia, Africa, the Caribbean and Latin America.

As part of IDRC's mollusc culture program, we supported a training course at Dalhousie University in 1982 and 1983. The emphasis was on the development of research skills to understand the biology of indigenous species and the adaptation and evaluation of culture methods. The course was not continued as it served the immediate training needs but it was decided that the lecture material developed for the course should be published to serve ongoing regional needs. These lectures have served as a basis for the present book.

The publication of this book, which has turned out to be a much expanded version of the original lecture notes, represents a significant step in IDRC's program to develop support systems for the development of small-scale bivalve farming in less developed countries. IDRC welcomes the opportunity to join with the World Aquaculture Society in publishing this book. This will extend the distribution of the book to the global aquaculture community. The basic principles presented here should be applicable in any setting. It is our hope that this material will be useful to researchers and bivalve farmers throughout the world. By providing in one volume information on a wide variety of culture methods it should stimulate thought and new culture ideas. Perhaps it will also stimulate others to share in the development of culture systems in the scientifically less developed countries with the application of the scientific method to serve the poor rural communities.

F. Brian Davy
Associate Director (Fisheries)
Agriculture, Food and Nutrition Sciences Division
International Development Research Centre

TABLE OF CONTENTS

INTRODUCTION

Interest in farming bivalves exists in many countries. In the temperate regions there are well established culture industries. New industries, such as clam and mussel farming in North America and oyster culture in Australia, have also become established in recent years. In the tropics there are a few bivalve culture industries; many have been established for quite a long time.

Unlike some forms of aquaculture such as the farming of Atlantic salmon or penaeid shrimps, the culture of bivalves does not appear to be suitable for rapid expansion. The one species that seems to be suitable for development in new places is the Pacific oyster, *Crassostrea gigas*. One of the features of the geographically widespread success stories of Pacific oysters, shrimps and salmon is that each deals with a single or similar species which have been found suitable to a number of different locations. Furthermore, they are cultured with a common technology suited to the particular biological characteristics of the species. Shrimp and salmon culture technologies are expensive and beyond the capability of small farmers.

One rarely finds new developments in the culture of bivalve species previously not cultured. Yet there are many species of bivalves, particularly in the tropics, that would appear suitable for culture. The main problem is that the biology of most of these species is poorly known. There are a variety of culture technologies that could be adapted to suit these species, but studies are required to understand the biology, to evaluate the culture techniques and to adapt the techniques to the species and local conditions.

It is the purpose of this book to provide in one source the basics of bivalve biology, alternative culture techniques and guidance in the implementation of the research and development work needed to culture bivalves. It is hoped that this will be the start for training those who will undertake this task. This book is not intended to be a complete reference, and readers are encouraged to use the literature and other sources for more in–depth information. A short bibliography is provided.

The text has been developed from lectures presented during bivalve culture training courses given at Dalhousie University under the sponsorship of the International Development Research Centre, Canada. The courses were to provide the basics of bivalve culture to some of those involved in bivalve culture development projects, mostly in the tropics. Although the focus of this book is to a certain extent on the tropics, extensive use is made of temperate bivalves as examples because of a lack of information about tropical bivalves.

A previous publication, *Tropical Oysters* (Quayle, 1980) dealt with oysters only. The present text expands the discussion of oyster biology and culture given in that book considerably. This text also integrates aspects of oyster and other bivalve culture which have been presented in separate publications by Quayle and other authors. The oyster, the most widely cultured bivalve, has been selected as the example on which to base discussion of the biology of the bivalve molluscs. Differences between oysters and mussels, clams and scallops are dealt with in sections pertaining to those species.

We hope this text is useful as an introduction and guide to the culture of bivalves and in directing the development of experimental systems. There is a tremendous amount of work to be done in adapting old culture methods to new species or in developing new methods. We have emphasized what we feel are the important subjects and methods. Experience has shown them to be valuable.

CHAPTER 1

BIVALVE BIOLOGY

In this chapter, the anatomy, breeding and some other aspects of bivalve physiology will be discussed briefly. A number of terms are introduced that are important for the biologist but less important for the culturalist. For more detail on the biology the reader should refer to a text on bivalve biology (see the Biblography). Following this an overview of the taxonomy of bivalves is presented. Again, the presentation is only a brief orientation.

ANATOMY (OYSTER)

The anatomy of the oyster is described to give the general bivalve anatomy. More detailed descriptions are available in Quayle (1969) and Galstoff (1964). Following this the major differences between oysters and mussels, clams and scallops will be presented.

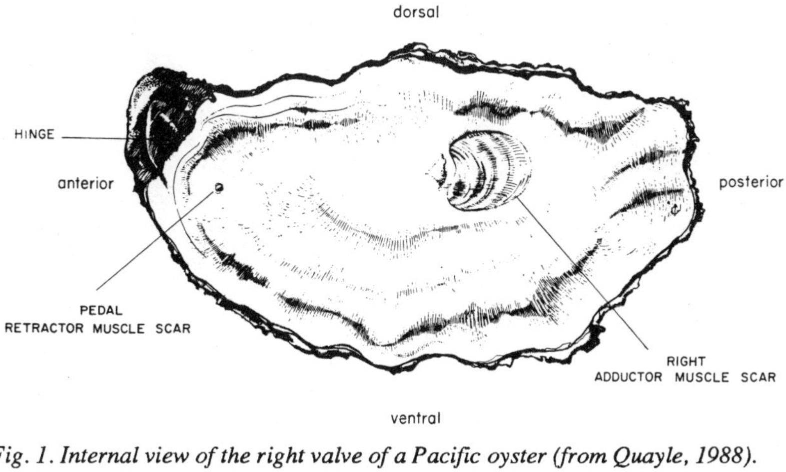

Fig. 1. Internal view of the right valve of a Pacific oyster (from Quayle, 1988).

Shell

The oyster consists of a soft body enclosed by a hard shell of two valves. The larger, lower valve (Fig. 1), usually cupped, is the left valve and is normally the area of attachment to the substrate. The upper right valve is smaller and usually quite flat. The shell is pointed at one end where the two valves are joined by a hinge ligament. The apex of the shell is in the form of pointed or rounded knobs called umbones (singular umbo). The ligament is of horny conchiolin. It is internal and tends to spring the valves apart.

The shell is composed of three layers (Fig. 2). The periostracum is a thin outer layer which is seldom apparent in oysters, for it is rapidly worn away. The periostracum is much more prominent in mussels. The central layer, which is the thickest of the three, is chalky in nature. The inner shell layer, next to the soft body is thin, often shiny or lustrous and very hard. This is called the nacreous layer or nacre.

The shell is composed primarily of calcium carbonate. The central layer and the nacre have different crystaline structures, giving different appearances and texture. The calcium carbonate of the shell is imbedded in a protein matrix. The hard part of the shell is only about 5% organic. The periostracum is almost all protein.

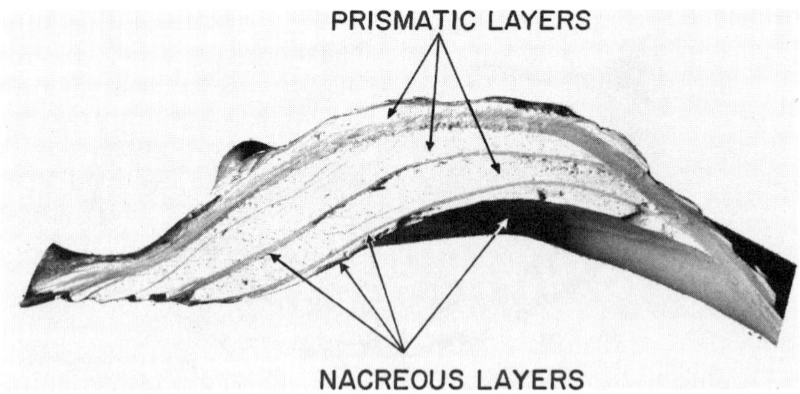

Fig. 2. Shell of a Pacific oyster cut through the hinge to show layers (from Quayle, 1988)

The shape of the oyster shell is extremely variable, depending on the environment in which the oyster is grown. If the oyster is grown individually on a soft bottom, the shell tends to be smooth and elongated. If grown on a hard bottom, such as gravel, the surface of the shell tends to become fluted or corrugated, the shape becomes more circular and the lower (left) valve becomes deeper. Proximity to other oysters, as when grown in a cluster, will cause great distortions in shape. If grown individually attached to a firm surface, the lower valve will grow attached to that surface rather than having the edge grow away from the surface, thus, losing the normal cupped appearance. This process is called xenomorphism (having the shape determined by contours of the substrate) and is an important factor in oyster culture. Waters of high salinity produce a harder shell than low salinity waters.

Body

With the shell lying on the left or cupped valve and the hinged or umbonal end directed to the right and the upper right valve removed, the body of the oyster lies with the mouth at the umbonal end (Fig. 3). Thus, the long axis of the oyster is actually the height (anterior-posterior axis), but common usage indicates it as the length. The dimension perpendicular to the height

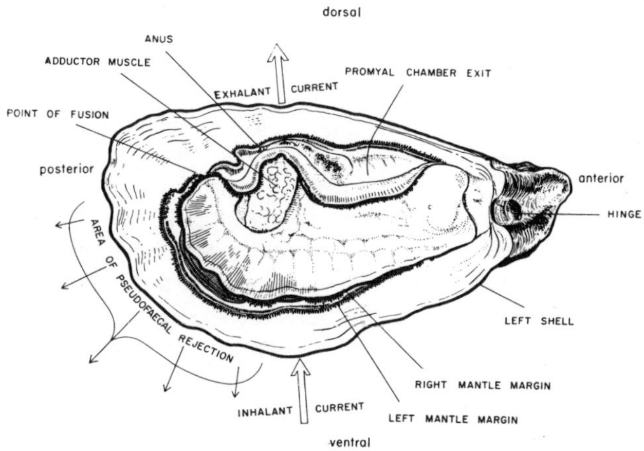

Fig. 3. Pacific oyster with right valve removed (from Quayle, 1988).

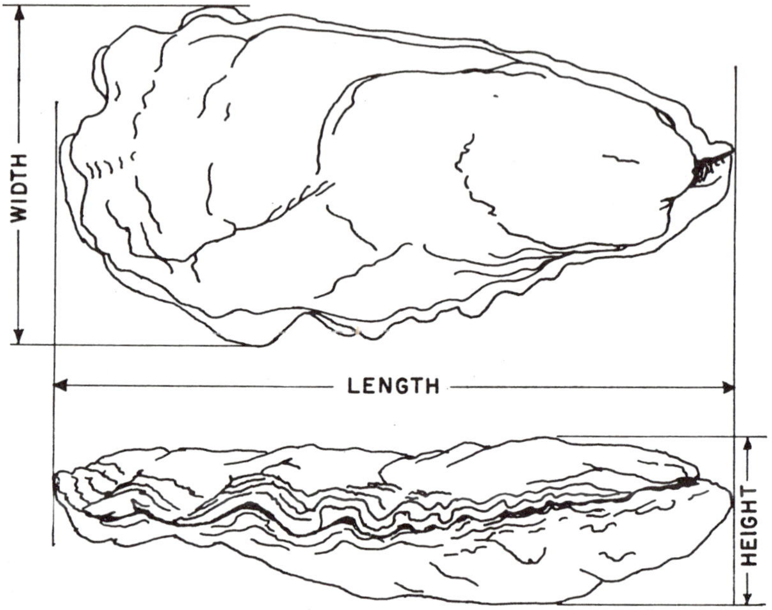

Fig. 4. Diagram to represent dimensional terms applies to oysters (from Quayle, 1988).

in the plane of the shell is the length (dorsal-ventral axis). Again common usage is different. It is referred to as the width. Thus, there is sometimes confusion. The dimension perpendicular to the shell is the depth (the thickness in common usage). To summarize, in lay terms, the anatomical dimensions height and length are called the length and width, respectively (Fig. 4).

About two thirds of the distance from the umbone is the adductor muscle which links the two valves together. It works continuously against the spring pressure of the hinge ligament which pushes to open the valves. The adductor muscle consists of two sections, a small crescent shaped section, pure white in colour, and a larger, anterior, beige coloured area. This large section is responsible for the main opening and closure of the valves and is called the "quick" muscle. The small white section is the "catch" muscle, for it can hold the valves in a set position for long periods with little expenditure of energy.

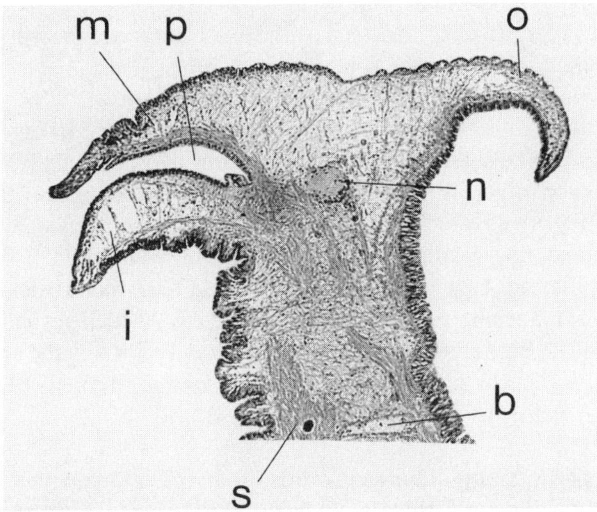

Fig. 5 Cross section of the mantle edge of the Pacific oyster. x 35. b = blood vessel; i = inner lobe; m = middle lobe; n = nerve; o = outer lobe; p = periostracal groove; s = muscle (from Quayle, 1988).

Covering the soft body, on each side except for the adductor muscle area, is the mantle. This is a thin skirt, thickened only at the edges and is the part of the body that deposits the shell. The thickened edge of the mantle is composed of three lobes (Fig. 5) that are mainly sensory and assist in controlling the inflow of water to the body. The mantle lobes are usually dark in colour from black to shades of brown, depending partly on the species and environment. In some species the lobes vary in colour; in others they are typically one colour. Over the mouth area two lobes of the mantle are fused to form a hood. At the posterior end, the two lobes are again fused at a single point and this separates the inflowing from the outgoing current (Fig. 3). In the genus *Crassostrea,* the mantle in front of the adductor muscle is not fused and forms the promyal chamber, a structure not found in the genus *Ostrea.* This chamber is an additional discharge area for outgoing water.

The mantle edge lays close to the edge of the shell where most of the shell is deposited. This is how the shell increases in length and width. Growth in shell length and width originates from the outer surface of the outer fold of the mantle edge. The inner surface of this fold secretes the periostracum. The inner shell surface (nacre) is formed by the outer surface of the whole mantle, thus creating thickness of the shell.

With the mantle lobes removed, four long finely ridged, beige coloured appendages can be seen on the lower or ventral part of the body. These are the ctenidia or gills (Fig. 6) which are the food collecting and respiratory organs. They are covered with rapidly beating fine hairs called cilia which create an incoming current of water through the ventral shell gape. The ctenidia (gills) are basket-like structures containing blood vessels. Food particles and other material in the water current are strained out and the blood is aerated. The filtered water is passed out dorsally through the area behind the adductor muscle and, in the genus *Crassostrea,* through the promyal chamber as well.

Food particles are directed towards the mouth along pathways on the gills by action of certain gill cilia. In front of (anterior to) the ctenidia (gills) is an additional set of four structures called labial palps (Figs. 6 and 7). It is

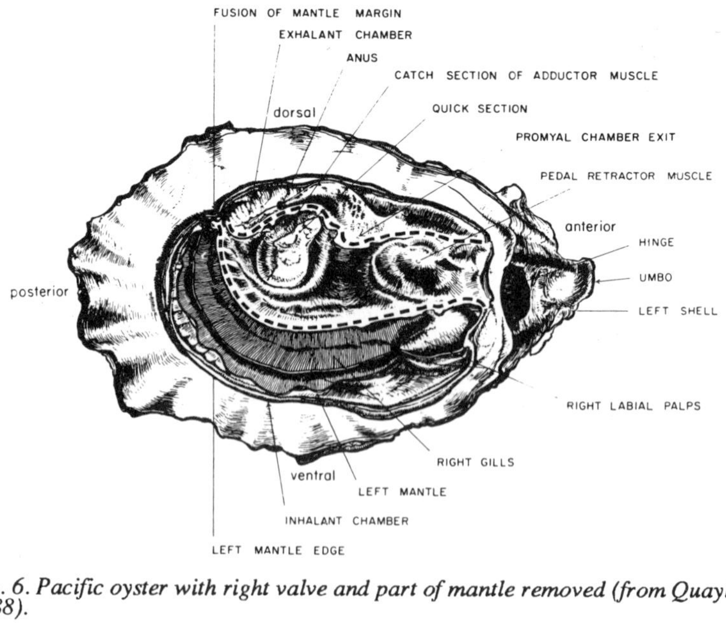

Fig. 6. Pacific oyster with right valve and part of mantle removed (from Quayle, 1988).

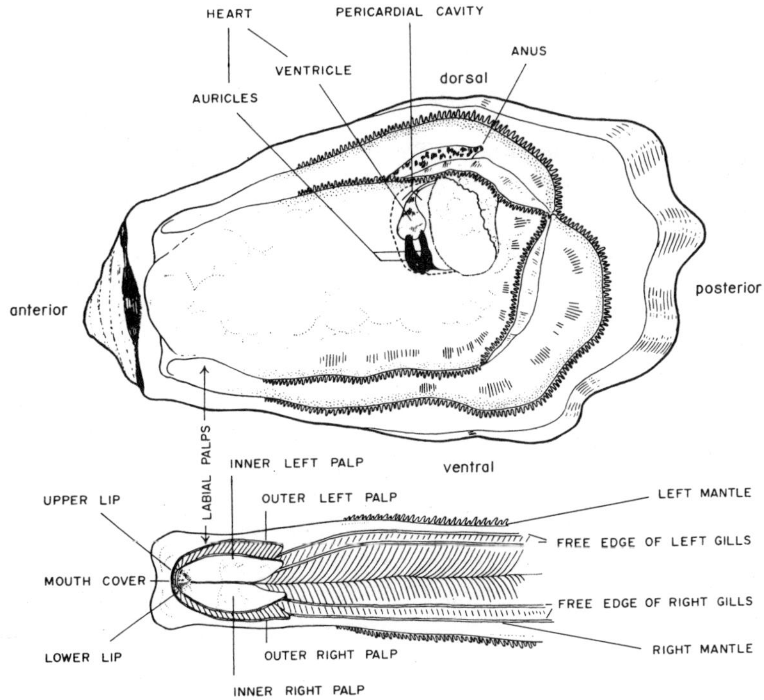

Fig. 7. (Top) Pacific oyster with the left valve removed and part of mantle cut away to expose the pericardial cavity. (Bottom) A ventral view of the mantle cavity to show the relationship between the palps and gills (from Quayle, 1988).

believed that the palps sort the food particles. From the labial palps the food particles enter the mouth under the oral hood. A short oesophagus leads into a stomach connected by tubes on either side to a network of tubules called the digestive diverticula, often referred to as the liver. This is the dark area on either side of the body, dark green or black when there is plenty of food in the water and light brown when food is scarce.

Opposite the oesophagus is the entrance to the intestine (also called the mid-gut), a narrow tube that loops around the stomach to end in the anus, which is located in the cloacal chamber just above the adductor muscle (Fig. 6). Close to the entrance to the intestine and associated with it is another narrow tube that has a blind end and is 2 to 3 cm long in an oyster of average size. This is called the crystalline style sac, which produces an extraordinary structure called the crystalline style, a gelatinous rod. It is golden brown or yellow in colour, flexible and tapered. Many mistakenly believe this to be a worm that infects the oyster. The crystalline style projects across the middle of the stomach to the opposite side where it rests against a hard transparent pad called the gastric shield.

While it has generally been considered that the style is rotated by cilia in the style sac, there are recent studies that indicate there may be little if any rotation in adult oysters. A thin layer of digestive material is moved along the style by action of the style sac cilia. This brings about contact between style and the absorptive areas of the style sac. The style itself is formed of concentric layers of mucoprotein. The style contains digestive enzymes that convert starch into sugars. Dissolution of the style in an actively feeding oyster occurs at the gastric shield where the acidity of stomach liquids is lower than in the region of the style sac. Dissolution of the whole style also occurs soon after an actively feeding oyster is removed from the water, which is why the style is so seldom observed. However, if the live oyster is returned to the water and allowed to feed the style is reformed in about an hour. It is an integral part of the digestive system.

Contained in the crystalline style sac of oysters (and other molluscs) are spirochaetes (*Cristispira*), which are elongated motile bacterialike organisms of microscopic size. Their function in the style area is not known, but they are harmless to oysters and to man.

In addition to the style sac, there are other shallow blind sacs or pouches in the stomach called caeca, which are concerned with food sorting. The course of food and digestion in the oyster stomach is a very complicated process and the following is a simplified account.

Mucus-enmeshed particles entering the oesophagus first meet the head of the crystalline style which mixes the food particles with enzymes it releases in this area. These enzymes effect extracellular digestion of starches. By a combination of ciliary pathways in the sorting pouches and in the stomach itself, small and partly digested particles are carried to the tubules of the digestive gland where intracellular digestion of fats and protein takes place.

In addition to these modes, digestion is also carried on by blood cells which are able to migrate in and out of the stomach and, by an engulfing action similar to that used by an amoeba, are able to ingest individual particles of food. Once inside the blood cell, the particles undergo intracellar digestion.

The caeca, in addition to directing particles destined for the digestive diverticula, also direct unwanted material along a special path, called a typhlosole, to the opening of the intestine. Waters from the digestive tubules are also carried to the intestine by this and other pathways. The main function of the intestine is to compact waste materials into solid strings and to carry them, via the anus, to the cloaca whence they are carried outside the shells by the exhalant water current.

Just anterior and close to the adductor muscle is the heart, enclosed in a chamber called the pericardial cavity (Fig. 7). The heart consists of a single muscular ventricle and two auricles. In a newly opened oyster the pulsations of the heart may be seen. Oxygenated blood comes to the heart from the gills and is sent through large arteries to small vessels. There are no small capillaries as such and most organs are simply bathed by the colourless blood. Deoxygenated blood is carried in poorly delineated veins to the gills for reoxygenation or to the kidneys, a small pair of tubes located under the adductor muscle and difficult to find. The kidneys discharge into the exhalant chamber.

Equally as simple as the circulatory system is the nervous system. There is no need for complexity because after the larval period the oyster does not move. In the adult oyster there are only two pairs of ganglia as compared to three in the larva and more motile bivalves such as clams. The pleural pair is near the mouth and the larger visceral pair is located under the adductor muscle. Fine nerve fibres enter the ganglia from various organs such as the mantle lobes and pass them on to the effector organs, the adductor muscle and glands.

The reproductive system consists of a series of branching tubules covering each side of the body and that end up in a single tube in the exhalant chamber. During the breeding season most of the body mass is composed of reproductive tissue and the discharge tubules may be seen as veins on the surface of the body. In the temperate zone, during the non-breeding season the gonadal tissue is replaced by another material (called Leydig tissue) consisting largely of glycogen. The germinal tissue is so reduced in size that the sex of the oyster is impossible to determine. In the tropics this period is of short duration because of the extended breeding season.

BREEDING

In the genus *Crassostrea* the sexes are separate. The female gonad or ovary produces eggs (ova). The male gonad or testes produces sperm. In both cases the gonadal products (sperm and eggs) are discharged externally into the open water, where fertilization takes place. The genital pores, through which eggs and sperm are discharged, exit into the exhalent chamber above the gills. Sperm is discharged through genital pores leaving the oyster with the exhalant water in a steady stream. In the case of the female spawning act, the adductor muscle is relaxed so the valves are kept open but the exhalant openings of the mantle are closed. Most of the inhalant area except for a small aperture in the posterior ventral area is also closed. As the eggs are released they cannot be discharged through the exhalent stream because the aperture is closed, so they pass through the opening in the gills and collect in the inhalant chamber. Then the adductor muscle contracts rapidly, and the cluster of eggs there is forced out the inhalant side in a cloud through the small opening left along the mantle edge curtain. In this way, the eggs may be ejected 30–50 cm away, thus ensuring some dispersal and an increased opportunity for them to be fertilized by sperm from adjacent males. The spawning act continues with discharge of eggs at intervals of 5 to 10 times per minute. Spawning may be complete at one time or there may be short pauses of hours, or longer ones of days or weeks, partly depending on the species, condition of ripeness and environmental factors. An average sized *Crassostrea* female (10 cm in length) may contain up to 100 million eggs. The number of sperm in males is many times higher. The size of ova of most oysters when spawned is about 70 µm (0.07 mm) and the sperm head about 3 µm (0.003 mm) (Fig. 8).

Females of the genus *Ostrea* also discharge the eggs through the gill pores (ostia), but in contrast to *Crassostrea*, the eggs are retained within the inhalant chamber where fertilization takes place. Sperm from adjacent spawning males enters with the inhalant water current. The fertilized eggs are incubated in that chamber for about10 days before being released as half grown larvae. Species that incubate the larvae are called larviparous. Most species of *Ostrea* discharge less than a million eggs.

In *Crassostrea*, the sex of the oyster may change between breeding seasons, but in *Ostrea* the sex may alternate once or several times within one breeding season, depending on temperature and food conditions. In general, the proportion of males and females remains approximately equal in spite of these sex changes.

Of importance in oyster culture is a knowledge of the seasonal change in condition of the gonad, for this delineates the breeding season and seasons of optimum condition ("fatness" or edibility) relative to marketing. For an understanding of these changes, an explanation of some terms is necessary, particularly to help in reading the literature. The process of producing eggs

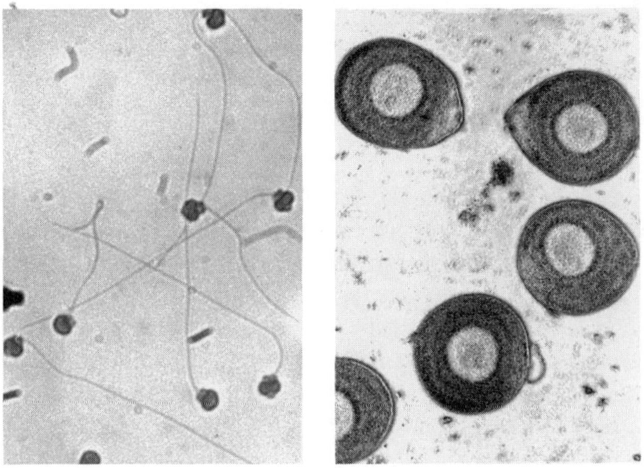

A B

Fig. 8. Photomicrographs of sperm (A. x1350) and egg (B. x270) of Pacific oyster (from Quayle, 1988).

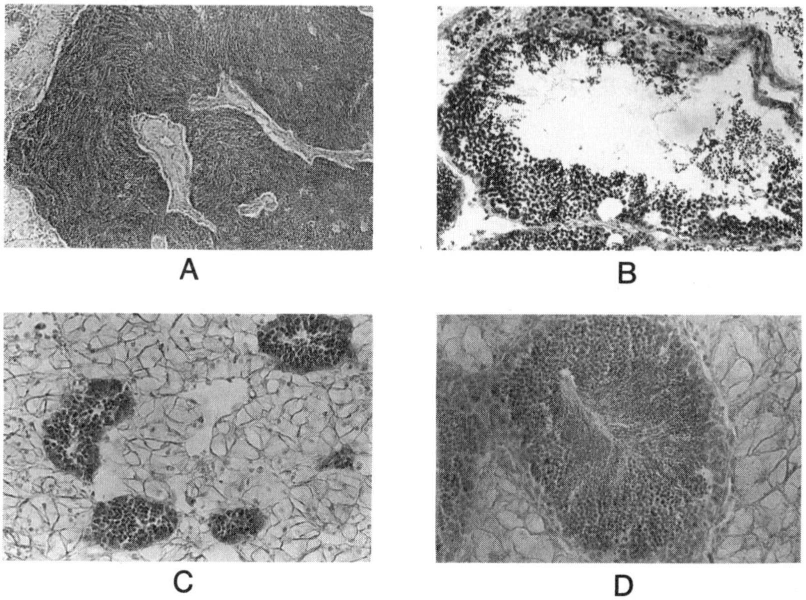

Fig. 9. Photomicrographs of sections of male Pacific oyster gonads to show seasonal changes. x69. A = ripe condition before spawning: B = spawned out condition with the follicle nearly empty. C = early development of male gonads: D = gonad approching ripeness with developing cells in the outer portion of the follicle (from Quayle, 1988).

or sperm (also called gametes) is gametogenesis: oogenesis for eggs and spermatogenesis for sperm. The actual cells are called gonocytes: ovocytes for eggs and spermatocytes for sperm. Secondary gonocytes follow the initial primary ones but in the case of spermatogenesis, the secondary spermatocytes are converted into spermatids and the spermatids into sperm. The normal seasonal gonadal cycle starts with a resting stage, which in temperate waters occurs during the winter. The gonadal material is so reduced it is not discernable and the sexes are undifferentiated. The tissue at this stage is composed of mainly glycogen, the animal starch which gives the oyster its unique flavour. This is called Leydig tissue.

In the temperate zone, gametogenesis begins when water temperatures rise in spring, about April and May. The gonad proliferates by enlargement

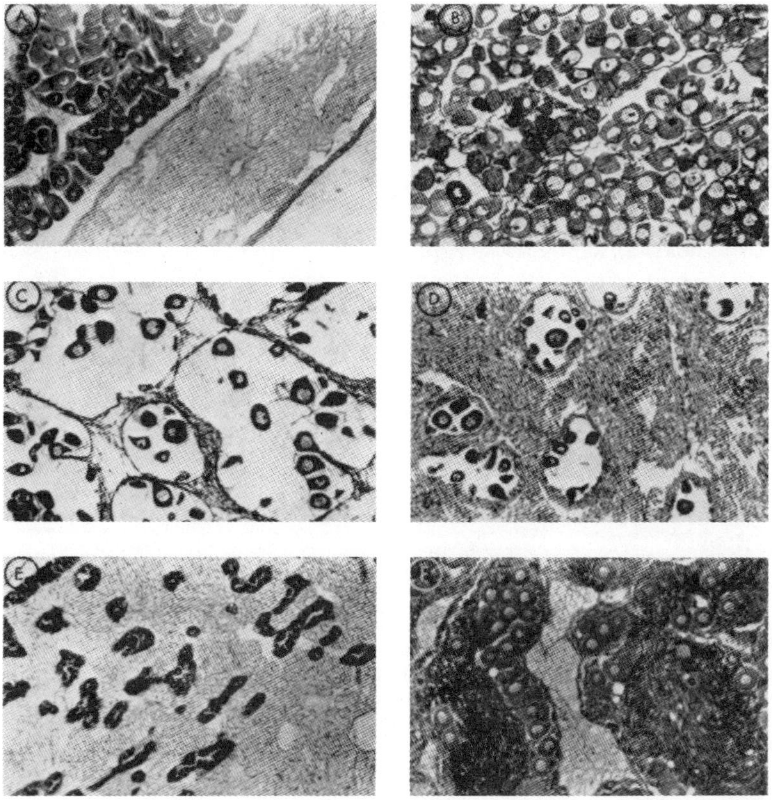

Fig. 10. Photomicrographs of sections of female Pacific oyster gonads to show seasonal changes. X68. A = ripe female with gonad covered with a thin layer of glycogen-rich tissue: B = fully ripe female: C = spawned out female with a few relict eggs: D = female in fall condition with follicles closing in on the relict ova: E = female in spring condition with early-stage developing eggs on follicle margins: F = hermaphrodite with both eggs and sperm (from Quayle, 1988).

of spaces called follicles. The follicles are lined with germinal cells which undergo the gametogenic cycle mentioned previously. As the follicles increase in size and become filled with gonocytes, the surrounding glycogen–bearing tissue becomes progressively reduced. In the fully ripe oyster, none of this material is left and the follicles are filled either with eggs or sperm and are ready to be spawned (Figs. 9 and 10). The glycogen–

gametogenesis cycle has been well documented in temperate oysters. There is very little evidence from the tropics, and the cycle is not as pronounced as in the temperate zone.

As with oysters, the breeding cycle of mussels, clams and scallops in temperate waters follows a seasonal cycle. However, there is not usually a resting period with a pronounced accumulation of glycogen. New gametes start to develop soon after spawning. The cycles of these bivalves has not been studied extensively in the tropics, but it would appear that they are similar to oysters in the tropics. That is, after the first maturation spawning is more or less constant unless there is a seasonal cycle imposed by salinity or productivity changes.

OTHER SPECIES

Mussels

The mussel has organ systems similar to oysters with some modifications. The mussel shell is typically pointed at the umbo which is terminal in *Mytilus* and *Perna* but slightly subterminal in *Modiolus* and several other species. The shell is rounded at the posterior end and has the same three layers as other bivalves (Fig. 2). Most mussels retain a dark periostracum. In some species, such as *Perna,* the green periostracum is very prominent. There are no true teeth in mussels.

Mussels have a foot like clams (Fig. 11), but it is much reduced in size. The foot is able to extend a considerable distance, providing limited mobility. An important part of the mussel foot is the byssal gland which secretes strong attachment threads. A mussel can detach these byssal strands and secrete new ones, enabling it to alter its position. When small, it is able to move by means of the ciliated foot.

The two adductor muscles are dissimilar in size, with the posterior one being the larger. This is called the heteromyarian condition. Another anatomical feature of mussels is the large retractor muscles, which are inserted in the foot. The posterior retractors, the larger of the two sets, are attached

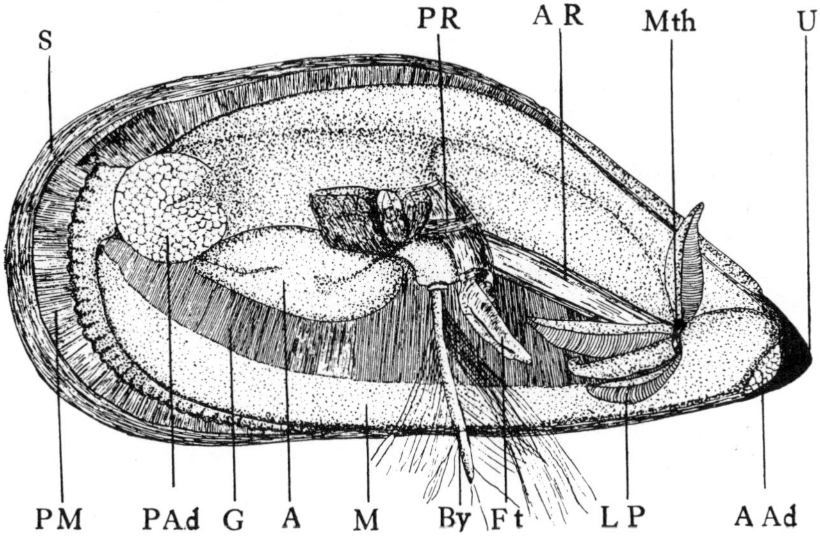

Fig. 11. Lateral view of a mussel with the shell and mantle of one side removed. Slightly enlarged. Abbreviations: A = abdomen; A, AD = anterior adductor muscle; AR = anterior retractor muscle; By = byssus; Ft = foot; G = gills; Lp = labial palps; M = mantle; Mth = mouth; PAd = posterior adductor muscle; PM = pallial muscles; PR = posterior retractor muscles; S = shell; U = umbo (from Field, 1911).

to the valves in the posterior dorsal area in front of the posterior adductor muscle. Anterior retractors are attached to the valves just above the small anterior adductor in the umbonal area.

A unique feature of mussels is the location of the gonads in the mantle, as well as in the body proper. Otherwise reproduction and development of planktotrophic larvae is similar to that of the oyster. The male gonad is creamy white in colour, while the female is pink or reddish. In the genus *Semimytilus*, the mantle on one side of the mussel contains testes while the other contains the ovary.

Clams

The basic anatomy, circulatory, nervous and sensory organ systems of clams are not too different from oysters. As would be expected from the large number of species, there is some diversity in detail in these systems but the main differences lie in the form of the shell, the presence of a foot and two adductor muscles in most but not all species. Some, such as the Tridacnidae, have deviated anatomically greatly from the typical edible clam form. The following anatomical description refers to a generalized clam common to most edible and harvested species.

The shell, as with oysters, has two valves. The shell material also consists of the three main layers characteristic of an oyster shell: an outer periostracum, a central prismatic layer and an inner nacreous layer. The periostracum in some species of clams, such as the Solenidae, is very apparent and maintains its identity throughout most of the life of the clam. In others, such as the Mactridae, it errodes in the umbonal region but is usually present along the shell edges.

If the two valves are of the same size they are called equivalve; if unequal in size they are inequivalve. If the umbones are centrally placed, the shell is equilateral; if displaced from the centre the shell is said to be inequilateral. There is often a gape in the shell to allow the protrusion of the siphon, and in the Solenidae such as *Solen* and *Ensis*, there are both anterior and posterior gapes, one for the foot and one for the siphon.

The exterior of some shells is smooth. Others are sculptured in various ways but mainly radially or concentrically and may be studded with ridges, spines, nodules, or leaf-like processes. Radial sculptures emanate from the umbones and run to the valve edge. Concentric sculptures are parallel to the edge of the valve. If there are both concentric and radial sculptures, the effect is said to be cancellate.

If the umbones point to each other they are said to be orthogyrate, if pointing to the posterior, opisthogyrate and if to the anterior, prosogyrate. In most clams umbones take the latter position with the ligament to the posterior. Immediately in front of the umbones, especially in the Veneridae, is a heart-shaped depression called the lunule. With the anterior end of the clam up and the hinge toward the holder, the valve to the left is the left valve. Another way in which to determine this is to observe the interior of the shell with the hinge directed away and the pallial sinus opens to the right of the right valve.

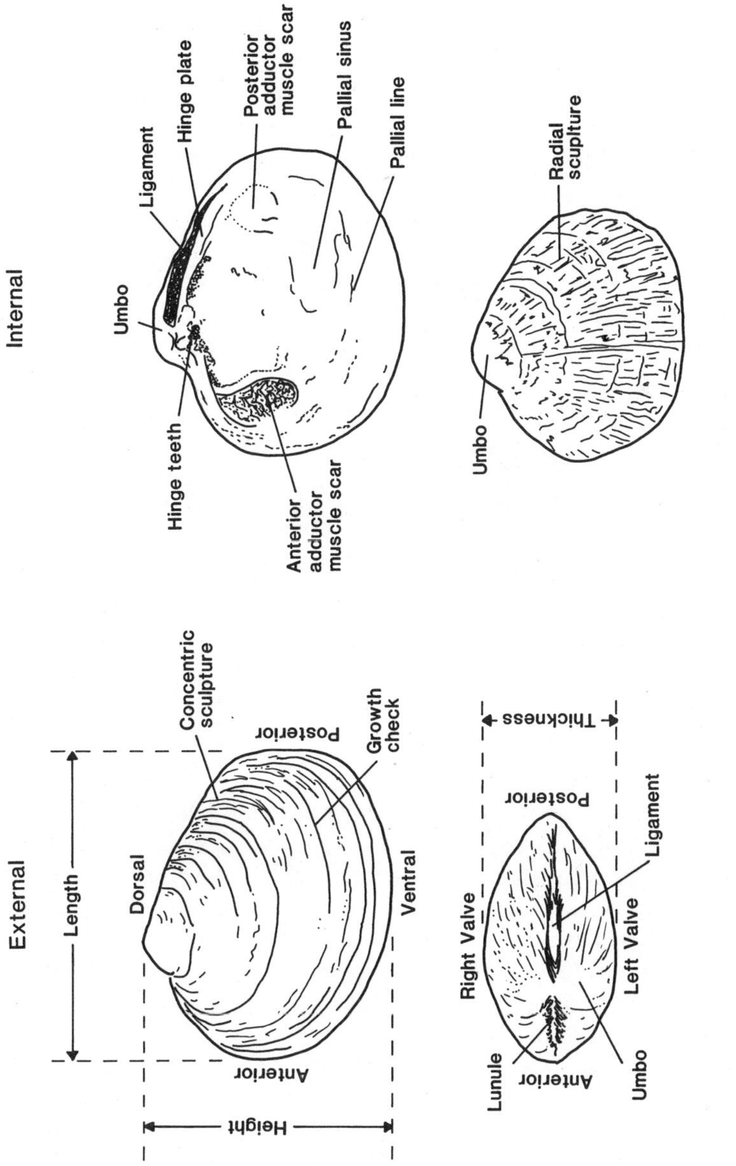

Fig. 12. Clam shell sculpture (redrawn from Quayle and Bourne, 1972).

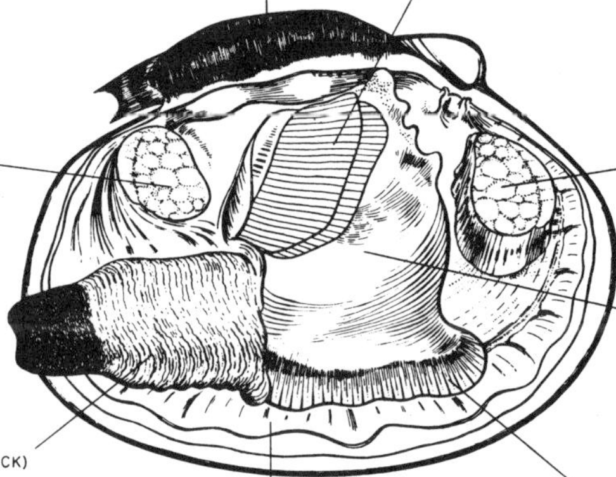

Fig. 13. Butter clam (Saxidomus giganteus) *with right valve removed to show main body parts (from Quayle and Bourne, 1972).*

Internally, the concave surface or nacreous layer usually has markings or scars that mark the attachment of muscles (Fig.12) and they are useful identification features. The main ones are the adductor muscle scars, that of the pallial sinus, which marks the presence and position of the siphon, and the pallial line, the area of attachment of the mantle edge. The anterior end of most clams is narrower than the posterior. Dorsally the valves are joined by a hinge ligament which is more conspicuous in clams than in oysters.

The hinge is usually brown in colour and chitinous in nature and springs the valves apart. Shape and size of the hinge, usually posterior to the umbones, vary with the species. In some clams the hinge may be internal or partially so. Also, in some species, in addition to the ligament, there may be an internal cartilage called the resilium. This is more elastic than the ligament and sits in a spoon–shaped projection of the valves just below the umbo called a chondrophore or resilifer. This structure occurs in the Mactridae and Myidae.

Internally and below the umbones are hinge teeth (interlocking shell projections) which vary in number, shape and position and are important as identification features. The teeth immediately below the umbones are cardinal and the ones on the side are laterals. However, some species are edentulous, having no teeth, while others (*Anadara*) have numerous chevronlike teeth.

The body of most clams varies from that of an oyster in the possession of two adductor muscles, maintaining the number occuring in the larvae (Fig. 13). In addition, the foot, which exists in the larva, is retained in the adult and takes on a variety of shapes and functions depending on the mode of life of the clam. The Solenidae, being active diggers, have a foot different from less active types. To maintain contact with the surface, the burrowers have a siphon or neck, which is a pair of tubes which can be extended to serve as a conduit for incoming and outgoing water. There is an inhalant siphon and an exhalant one, replacing the area system of the oyster. This modifies the mantle edge, from which the siphons are formed, and there are various degrees of mantle fusion according to species. In most Veneridae, there is little mantle fusion, but in the Hiatellidae, such as *Panope* spp, mantle fusion is almost complete. In this genus the siphon is able to extend almost 1 m. The muscular attachment of the siphonal apparatus is reflected in shell scars, termed the pallial sinus (Fig. 12), and the depth of the sinus is an indication of the size of the siphons.

The body wall of most clams tends to be more muscular than that of the oyster and this affects the appearance of the clam during the reproductive cycle. When an oyster is ripe, individual tubules may be seen as vein-like structures but these are not visible in a ripe clam. The ripe clam gonad is much more compact than the oyster's, and a ripe clam is at its peak in edibility as well as meat productivity. A spawned-out clam is usually dark in colour, mainly muscle and tough to eat. By contrast, the oyster is most palatable during the glycogenous period.

Scallops

Pectens are nearly all quite circular in outline and most have prominent radiating ribs. The hinge line is straight and terminates in ear-like projections called auricles. These are asymmetrical, the one on the right valve anterior being indented to form a byssal notch (Fig. 14). The ventral side of the byssal notch of the right valve has a number of small teeth called ctenolia (singular-ctenolium) which assist in aligning byssal threads and may be useful in the identification of small scallops. The umbonal hinge region is said to be dorsal, so with the hinge held upwards, and the byssal notch forward, the left and right valves are in the correct position. Contrary to the normal position of the oyster with the left valve down, scallops lie with the right valve down.

Valves may be equal or unequal. In some species the curvature of the two valves is identical; in others the left one may be quite flat. It is possible to age most scallops by growth checks which show up as concentric rings. The number and prominence of radial ribs, where present, are useful for species identification.

There is an internal and external part of the hinge ligament. The external is long and thin while the internal is short, thick and rubbery, like the resilium of the Mactridae, and also sits in a resilifer. Both parts of the ligament tend to spring the valves apart. Growth lines on the resilium have been used to determine age in some species.

The dominant feature of the anatomy of the scallop is the large single adductor muscle, which in most countries is the main edible part. Like all bivalves, the larval scallop has two adductor muscles but after settlement the posterior muscle enlarges and migrates to the approximate center of the shell (Fig. 15). It can form 30% of the whole meat weight and is a main center of glycogen storage in contrast to Leydig tissue in most other bivalves.

The scallop mantle is highly developed, as it is important in the swimming activity of this group. In addition, it is the location of numerous eyes which contain a cornea, lens, retina and optic nerve. With these eyes it appears the scallop can detect movement and react to light intensity changes.

The foot is relatively small and mainly concerned with byssus formation for it contains the byssal gland. Initially, at and immediately after setting, there is considerable byssal activity, but later it is lost in some species or reduced in others.

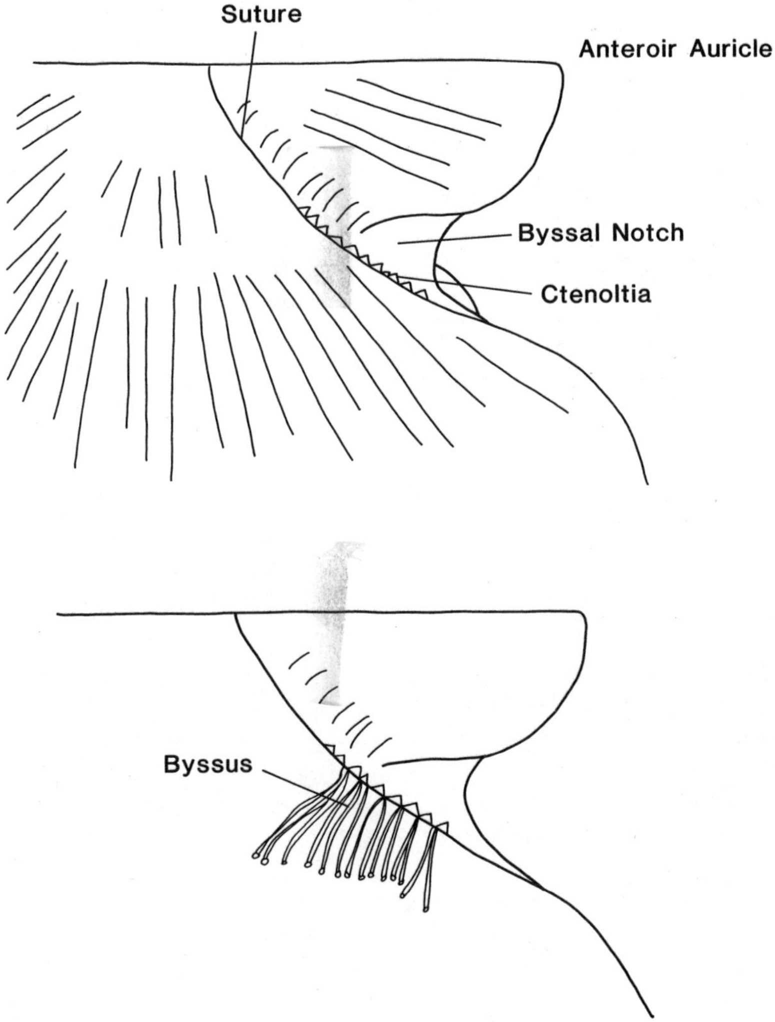

Fig. 14. Scallop ear anatomy.

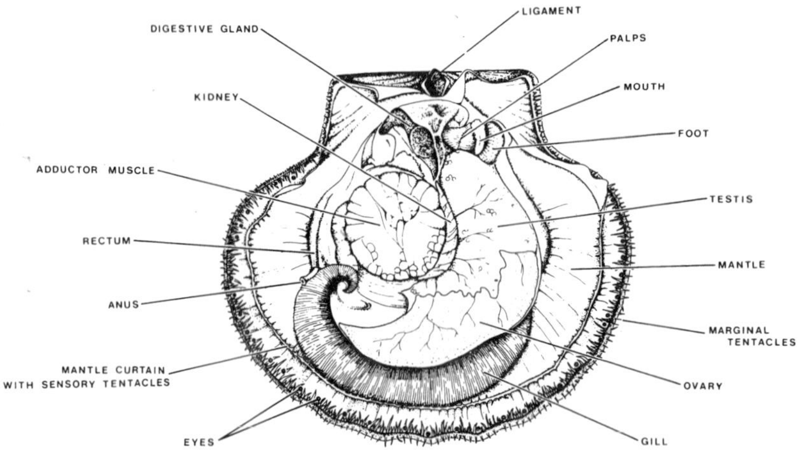

Fig. 15. Left valve Pecten irradians: *mantle and gills have been removed to show internal anatomy and part of the outer wall of the visceral mass dissected to make visible the digestive tract (from Mason, 1983).*

Some species of scallops are hermaphroditic, with both ovary and testis located in one gonad. The differences are conspicuous in ripe animals, with the posterior ovary pink or red in colour and the anterior testis white or beige. The same colour differences exist in species where sexes are separate. Otherwise the organs such as nerve, circulatory and digestive systems are basically similar to those in other bivalves.

Scallops possess a unique ability to swim, although one group (*Crassadoma* formerly *Hinnites*) loses this ability after a year or so and becomes cemented to the substrate. The swimming process is essentially one of jet propulsion. Power comes from rapid closure of the valves, forcing the water from the mantle cavity. The direction of the water jet results from the openings and closures of specific areas of the mantle curtain. The main swimming operation occurs with curtain openings at the ends of the hinge lines, and jets force the scallop with the ventral edge forward. Continued contractions of the adductor enables the scallop to move several metres in one swim.

An escape reaction stimulated by the approach of a predator, such as a seastar, causes movement in the direction of the hinge line. Other positions of mantle curtain opening and closures can also permit turning or twisting motions. In spite of the swimming ability of scallops, there is generally little movement unless caused by disturbances. There are no indications of large scale movement from one area to another in most species.

TAXONOMY

Oysters, mussels, clams and scallops belong to the Phylum Mollusca, a group that includes such diverse animals as chitons, snails, tusk shells, squid and octupos. The phylum Mollusca has six classes. Oysters, mussels, clams and scallops belong to the class Lamellibranchiata or Bivalvia. The name Lamellibranchiata refers to the gills (= branchia) which are characteristically lamellar in appearance. Gills are the principal organ of food collection as well as respiration of these animals. The group is also called the Bivalvia because of the two shells (= valves) which enclose the animals. They are also sometimes referred to as Pelecypoda or Acephala.

There are over 7000 species of bivalves that live in a wide variety of habitats from very shallow water to the deepest part of the ocean. They also take on numerous shapes and forms, some barely recognizable as bivalves. The principal forms of interest to us are those that are fished extensively and are, or potentially are, cultured. These are found in several families. The Ostreidae include the various oysters that are found throughout the world. The Mytilidae are the mussels. The scallops are in the family Pectinidae. There are a number of families that include the other species in which we will be interested. These are commonly called the clams.

Oysters

In spite of the intense biological study of oysters through the years, the taxonomy is still in a confused state. Relatively recently the cupped, oviparous oysters were removed from *Ostrea* to the genus *Crassostrea*, with the larviparous flat oysters retained in the genus *Ostrea*. (Larviparous species incubate the eggs inside the mother's shell for part of the larval development while in oviparous species all of the larval development occurs in the free swimming state.) Still more recently, other oviparous oysters, notably the Australian and New Zealand species, were placed in the genus *Saccostrea* while the quite different larviparous Chilean and another New Zealand oyster were assigned to a new genus *Tiostrea*. Another new genus now appearing in the literature is *Striostrea*, with a species in South Africa and one in the eastern tropical Pacific.

Representative of another large group of oysters (*Pycnodonta*) is *Hyotissa hyotis*, which, while quite widespread subtidally on tropical rock and coral reefs, is unlikely to become a commercial species. It is sometimes called the coxcomb oyster because of its deeply indented posterior margin.

Crassostrea are fairly large oysters, up to 200 mm or more. They are primarily brackishwater species with a wide salinity tolerance. The typical habitat is intertidal. Chomata are absent. The adductor muscle scar is reniform and variously coloured according to species. Chalky deposits are often present in the shell. This genus has a promyal chamber.

Ostrea species are moderate to large in size with a flat right valve. They generally tend to have less tolerance for low salinity than *Crassostrea* but are found in estuarine habitats. Chomata are present near the hinge, but are small and inconspicuous. Small radial plicae are found in the left valve. The nacre (inside layer of the shell) is often coloured, the muscle scar reniform, and there is no promyal chamber.

Saccostrea are medium sized oysters with the left valve plicate. The salinity tolerance and habitat are similar to *Crassostrea*. Chomata are present, often completely around the periphery. The muscle scar is reniform and generally coloured. As in *Crassostrea*, *Saccostrea* species have a promyal chamber.

Tiostrea species are medium in size. Chomata are present in young animals but may disappear with age. The adductor muscle scar is reniform.

Striostrea are large up to 200 mm. They are characterized by full left valve attachment. The right valve has brittle lamellae. The chomata are large and elongate. They occupy shallow subtidal habitats.

Hyotissa are large, up to 250 mm. The valves are thick. A promyal chamber is present. The posterior margin is deeply folded.

The main commercial species likely to be encountered in the various regions are as follows. These taxonomic designations are based largely on local usage and not to be taken as the last word on correct nomenclature.

Crassostrea gigas - Japan and southward; transplanted throughout the world although not particularly successful in the tropics.
Crassostrea rivularis - Japan
Crassostrea belcheri - South East Asia
Crassostrea iredalei - South East Asia
Crassostrea madrasensis - India, Sri Lanka
Crassostrea (Saccostrea) echinata - South East Asia
Crassostrea virginica - eastern North America from Canada south to the Gulf of Mexico
Crassostrea corteziensis - Gulf of California to Panama
Crassostrea rhizophorae - Caribbean Islands and adjacent South America
Crassostrea gasar (=tulipa) - West Africa
Crassostrea brasiliana - Southeastern South America
Crassostrea paraibanensis - Northeastern Brazil
Saccostrea cucullata - South East Asia and Indian Ocean
Saccostrea glomerata - New Zealand
Saccostrea commercialis - Australia
Saccostrea (Crassostrea) lugubris - Thailand and South China Sea

Striostrea margaritacea - South Africa
Striostrea prismatica - Western tropical South America
Ostrea edulis - Western Europe and Mediterranean
Ostrea lurida - Western North America
Tiostrea chilensis - Chile
Tiostrea lutaria - New Zealand
Hyotissa hyotis - worldwide, tropical

Mussels

Mussels belong to the family Mytilidae and are among the most familiar of all bivalves, for they are widely distributed throughout all oceans. This is a large group of molluscs which include a wide spectrum of bivalves, generally typified by the common mussel *Mytilus edulis*. There are about 75 mytilid species on the west coasts of North and South America alone, only six of which have potential for culture. The group has a large range in size and habitat, with the common characteristic of byssal attachment. Most species are intertidal but there are subtidal species. The habitats vary from rocks to gravel beaches and several species, e.g., date mussels, are able to burrow into soft rock, while others live partially buried in mud or sand. Many species are small, less than one cm in length, and one species spins a cocoon or nest of byssal threads. Some attain a length of up to 20 cm.

The main genus now being cultured in the tropics is *Perna* (Fig. 16), of which there are several species occuring in Venezuela, Singapore, Malaysia, India, the Philippines, and New Zealand. The dominant cultured species however, is *Mytilus edulis*, with the centre of production in western Europe, with an annual production of 420,000 t.

The genera found in the tropics are *Perna*, *Mytilus*, *Aulacomya*, and *Modiolus* (Fig. 16). The distribution of *Perna* throughout the world is somewhat odd, with commercially exploitable populations in Venezuela, South Africa, New Zealand, India and specific parts of Southeast Asia such as Thailand, Malaysia, Singapore, and the Philippines. The main species are *Perna perna* (India, S.E. Asia, Venezuela and South Africa), *P. viridis* (India and S.E. Asia) and *P. canaliculus* (New Zealand).

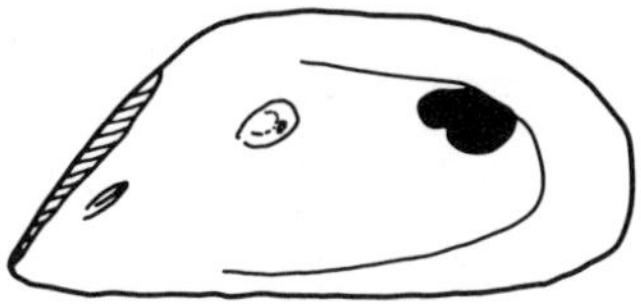

Perna perna

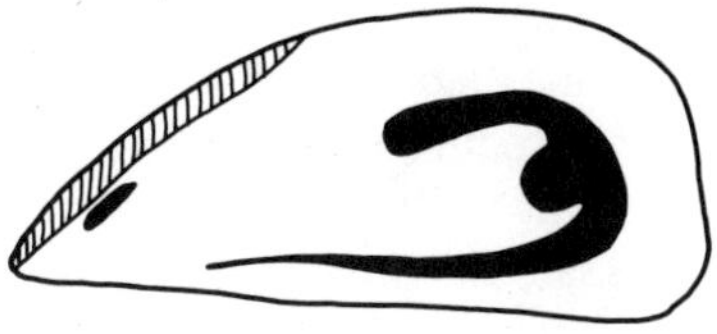

Mytilus edulis

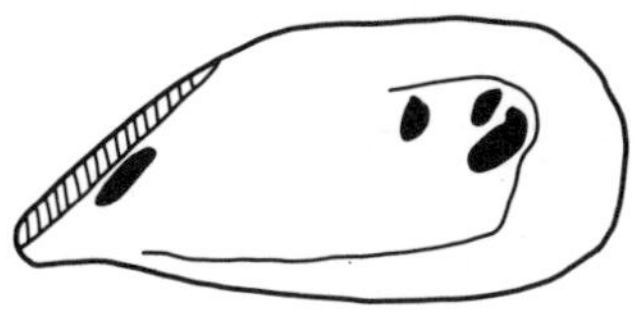

Perna viridis

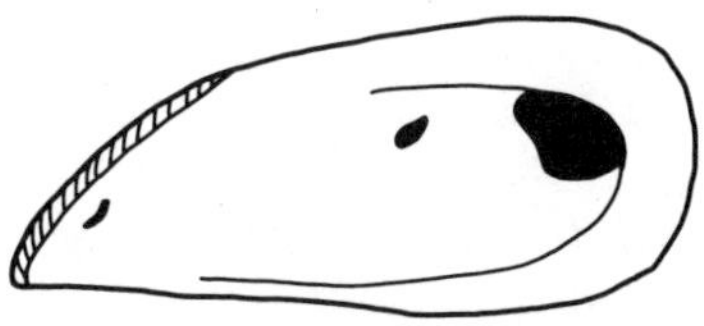

Perna canaliculus

Fig. 16. Sketches of the right valves of four mussel species.

Mytilus species are small to large. The shell is smooth or slightly ribbed in some species. The anterior adductor muscle is present with one retractor scar. There may be one or two teeth but no laterals. The shiny, smooth periostracum is black or brown (Fig. 16).

Perna species have no anterior adductor muscle and two adductor muscle scars. There are 10 to 18 primary lateral teeth and most often two, but sometimes only one, dysodont tooth. The shell is smooth with a pitted resilian ridge.

The shell of *Aulaçomya* species is strongly ribbed with a compact resilian ridge. There are no teeth, one retractor scar. The anterior adductor is absent in adults.

Modiolus species are usually brown with a thick periostracum, often with hair. Posterior retractors are continuous.

Clams

While there are many local names for the various species, the term clam generally includes all those species of bivalves not considered to be an oyster, mussel or scallop. The important clam families include the Veneridae, Mactridae, Solenidae, Myidae, Arcidae, Tridacnidae and Cardiidae. Species in the latter families are often called cockles being characterized by deep radial ribs.

Clams are identified by shape, colour, surface sculpture, muscle scars on the interior of the shell and particularly the hinge teeth (Fig. 12). The Veneridae, carpet or venus shells, range in size from a few millimetres to 10 to 12 cm in length with solid, equivalve shells. They are active burrowers mainly in firm substrates from the intertidal to moderately deep water. *Saxidomus*, *Amighinomya*, *Protothaca*, *Mercenaria*, *Tivela* and *Tapes* are some of the genera for which there are fisheries, with some culture of *Tapes* and *Mercenaria*. The siphons take a variety of forms and lengths depending on the depth to which they burrow, which is from just below the surface down to 30 cm or more.

The Mactridae are fairly large deep burrowers with heavy siphons covered with layers of periostracum. They may be as large as 20 cm in length and burrow to the depth of 45 cm in sandy-mud substrates, both intertidally and in deeper water. Among the genera fished commercially are *Spisula* and *Tresus*. None of this family is cultured.

The Myidae are moderately deep burrowers in sand and mud soils and the genus *Mya*, the soft shelled clam, is found in most northern seas where it is fished. Attempts to culture *Mya arenaria* have not been successful.

The Solenidae or razor clams are fairly deep, rapid burrowers, mainly in sand in the intertidal area or in shallow waters. The genera *Solen*, *Ensis*, *Siliqua* and *Sinonvacula* are fished commercially but, except for the latter in China, have not been cultured for their habits do not lend themselves well to culture. *Ensis* and *Siliqua* spp. may attain a length of nearly 20 cm.

The Arcidae or ark shells are almost all tropical and inhabit quite soft muddy ground. The two main genera are *Arca* and *Anadara*. Some have quite massive shells but others such as *Anadara granosa* are quite small (3 cm) and light. The latter occurs in great abundance in the area of Malacca Straits of Malaysia where it is known as the blood cockle and is widely cultured.

The Cardiidae are the cockles, not to be confused with *Anadara*. They are generally shallow burrowers living mainly on sandy bottoms, sometimes right on the surface. Some species attain a length of 15 cm. There is extensive use of cockles in Europe, but no culture.

The second family of cockles is the Tridacnidae, the giant clams. These are completely tropical and some species attain a great size, to 137 cm in length with two valves weighing up to 230 kg. They mainly inhabit coral reefs and are fished extensively for the highly valued adductor muscle. Experimental culture is now being conducted.

Scallops

Scallops, also called clams in Great Britain or Coquille St. Jacques in Spain and France, belong to the family Pectinidae in which there are several hundred species. Many are quite small, about 10 mm in diameter, while the largest measure up to 30 cm in length. The number of commercial scallops number only a dozen or so. The main commercial genera are *Pecten*, *Placopecten*, *Patinopecten*, *Aequipecten*, *Argopecten* and *Chlamys*.

Shells of *Pecten* species are large. The ribbed right (lower) valve is inflated while the upper valve is usually flat but may rarely be concave. The genus includes both hermaphrodites and species with separate sexes.

The valves of *Placopecten* are without ribs but do have radial striae (fine lines). However, the right valve is smoother. Both valves are slightly convex. Sexes are separate.

Shell of *Patinopecten* are large and circular. They have weakly ribbed valves which are nearly equal in convexity. A byssal notch is prominent and deep. Sexes are separate.

Upper valves of *Aequipecten* species are convex while the right valves are flattish. The valves are usually gaping.

In *Argopecten* both valves are convex but the lower one is more convex. The ears are equal in length. Radial ribs are strong and squarish. The species are all hermaphrodites.

The valves of *Chlamys* have unequal ears and a deep byssal notch. The valves are elongate, rounded ventrally and have strong primary ribs. Sexes are separate.

CHAPTER 2

SEED PRODUCTION FOR BIVALVE CULTURE

Clearly one of the essential elements for the culture of any bivalve is a source of seed stock. There are two basic sources of seed for culture: wild and hatchery stock. The former is produced from naturally spawning of stocks in the field. These stocks may be wild stocks or existing cultured stock. In any case the essential feature is that the control of reproduction is under natural forces. (In some circumstances, it may be possible to induce spawning in natural populations.) Hatchery seed, on the other hand, is from controlled production, although there are, in fact, varying degrees of control possible depending on the species. The availability of hatchery seed is increasing but is limited both in terms of species and geography. In this text we will spend little time on the production of hatchery seed.

Since seed supply is so critical, it is essential that the bivalve farmer is able to determine the seasonal supply cycles in order to maximize the collection of seed. This will involve being able to monitor the breeding cycle of stock. There are various methods available for this which will be discussed. Understanding the spawning, larval development and subsequent attachment is important in order to be able to predict when and where to collect spat. Spatfall forecasting is then possible. Since natural forces control the cycle, it is often possible to relate the cycles to environmental variables. Understanding these and the requirements of the larvae for substrate will make the spat collection process more efficient. The proper choice and use of the substrate for collecting seed is critical for the economical operation of a farm. There are practical considerations that apply to each situation.

Laboratory and microscope techniques are important particularly in field studies of reproduction cycles, larval monitoring and seed supply. Some useful guides to preservation and preparation techniques are given in Appendix II.

BREEDING CYCLE

Study of Gonadal Changes

Seasonal gonadal changes may be studied by gonadal smears, histological preparations, biochemical composition, or condition factor and visual observation.

Gonadal smears

This involves making a small cut in the surface of the body of the oyster or clam with a scalpel at a point half way between the position of the mouth and the adductor muscle. This should always be done in the same place. In mussels this should be done on the mantle and in scallops on the separate gonad. The edge of the scalpel is then stroked lightly over a glass microscope slide and examined immediately under a microscope with a magnification of about x100. This method is fairly satisfactory for females. The presence or absence of eggs and their approximate stage of development may be determined. Specific stage identification is difficult with this method. However, with experience the development from round, pale eggs to pear shaped, dense eggs full of yolk can be recognized. Owing to the small size of spermatogonia it is difficult to determine male stages, but it may be assumed that the timing of male development will approximate that of the females. A sample of about 25 females per month is required. More frequent samples may be necessary near spawning time.

Histological preparations

Histological preparations provide the most exact assessment of the gametogenic state of the molluscan gonad, but equipment and expertise are not always available. The process consists of preserving a small portion of the gonad 1 cm x 1cm x 1cm approximately, in a fixing fluid, removal of the water so the tissue may be infiltrated with wax resulting in a solid block of waxed tissue. This is then sliced into thin sections (10 µm), stained and mounted on a glass microscope slide. The process requires a number of chemicals such as fixing fluid, alcohol, xylol, stains and mounting fluid, wax, glass slides and cover slips, in addition to a variety of glassware, an oven and a costly machine called a microtome for cutting sections. What is seen in the sections represents fairly closely the situation in the gonad as the state of development of individual cells can be seen (Fig. 9 and 10). Both a qualitative and quantitative estimation of the degree of development can be obtained. At least 25 animals per month are required. Anything less would be wasted effort, because the results may not be reliable. It is possible to have such histological work contracted out to properly equipped laboratories, if local facilities are not available.

Biochemical composition

The biochemical composition of oysters changes as gametogenesis proceeds. The stores of glycogen are used to produce gametes. Some of the glycogen is converted into lipid which is the energy reserve for the larvae. Thus, it is possible to follow the reproductive cycle with changes in the biochemical composition. However, this requires a well equipped chemical laboratory and trained chemists. The oysters are analysed for such components as fats, proteins and glycogen. For most projects in the initial stages such methods are not necessary. The percentage of solids (discussed below) in an oyster is usually highly correlated with the biochemical changes and is easier to use. This is the proportion of dry meat weight compared to wet meat weight. A typical analysis is shown in Table 1.

Condition factor

The lipid (fat) content of molluscs is very low. However, in lay terms bivalves are sometimes called "fat." This refers to plump animals, that is, ones that fill the avaliable shell volume. The annual cycle of "fatness" is described in terms of the plumpness or the extent to which the oyster body fills the shell cavity. This is referred to as the "condition" of the oyster and is determined by the glycogen content and amount of reproductive tissue. The changes in glycogen content or reproductive tissue is reflected in the relationship between the size of the oyster and the meat content. The true oyster flavour is derived from the glycogen content. While an oyster in the spawning condition is edible it is not palatable to some people having a flavour different from, and usually not as acceptable as, the glycogenous state. (However, in Australia they prefer oysters to be "spawny.") There are

	Solids calculated in moisture-free flesh						
	Average weight (g)	Moisture (%)	Protein (%)	Glycogen (%)	Fat (%)	Ash (%)	Balance (%)
Feb. 9	18.90	78.20	47.80	20.50	10.68	8.66	12.36
Apr. 12	18.30	79.94	46.65	24.95	12.94	7.62	7.84
May 30	16.72	77.50	47.50	23.80	13.31	6.87	10.32
Aug. 3	11.07	81.41	54.60	11.85	15.75	7.80	10.00
Oct. 3	17.40	80.00	52.20	14.25	11.72	5.78	16.05
Dec. 7	20.10	83.37	49.90	20.05	13.08	8.42	9.45
Feb. 6	20.00	79.80	49.60	19.00	15.27	6.81	9.32

Table 1. Chemical composition of Pacific oysters in British Colombia (source: Quayle, 1988.)

several ways of measuring condition. These are often measures of the meat content relative to the shell volume that can be potentially filled. Such a measure is called a condition factor.

In most countries of the northern hemisphere where oysters are eaten raw, there is prevalent an idea that oysters should be eaten only in those months with an "R" in the spelling, in other words, from September through April. This likely originated in Europe where the favourite is the European flat oyster, *Ostrea edulis*, the female of which, as has been explained, incubates the larvae during the summer breeding season, the "R-less" months. The incubating larvae have shells and when eaten, resemble sand. Consequently the oyster, while edible, is not palatable because of the gritty texture. This situation is not found in *Crassostrea.* After an oyster has spawned, often in the "R-less" months in northern temperate regions, it is thin and watery, unattractive in appearance, without flavour, and low in meat yield.

There are several ways to measure the condition of molluscs quantitatively. In some cases the quantity of meat (volume or weight) is compared to the volume of oysters in the shell. For instance if 2 litres of oyster meat are produced from 10 litres of oysters in the shell, the return is 20%. An increase in the percentage indicates an increase in the condition or "fatness" of the oyster. The mistake in refering to the number of oyster meats/litres as a measure of condition is often made. This is in error, unless the sizes of the oysters are indicated. The litre could be composed of meats of 25 large oysters in poor condition or of 25 small oysters in good condition.

A commonly used measure is a condition factor or condition index which compares the weight or volume of the oyster meat with the volume of the shell cavity. The weight may be of wet oysters meats, just as they are shucked and drained for a specific time, or when dried at 98 °C to a constant weight. One condition factor is:

$$\frac{\text{weight of dry meat}}{\text{volume of shell cavity}} \times 1000$$

and an index of 70 indicates a thin oyster in poor condition, while one of 150 indicates a fat oyster. The seasonal changes in condition factor in temperate waters is shown in Fig. 17.

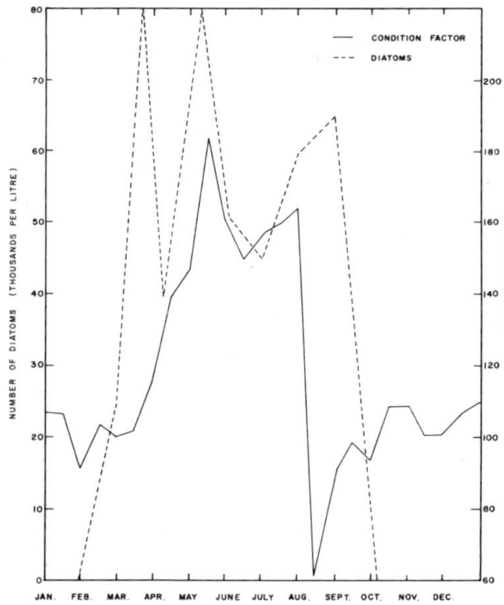

Fig. 17. Relationship between the standing crop (abundance) of diatoms and condition factor of Pacific oysters in Ladysmith Harbour (from Quayle, 1988).

One procedure for measuring the condition factor is as follows:

1. Select shells with unbroken edges.
2. Clean oysters of mud or fouling organisms.
3. Weigh oysters in air.
4. Weigh oysters in water (Fig. 18).
5. The difference between 3 and 4 is the whole volume.
6. Shuck and drain meats.
7. Weigh valves in air.
8. Weigh valves in water (Fig. 18).
9. The difference between 7 and 8 is the volume of the shell.
10. The difference between 5 (whole volume) and 9 (shell volume) is the internal volume.
11. Dry oyster meats in an incubator at 95 to 100 °C until the meat reaches a constant weight. The time will depend on the size of the oyster, but usually between 24 and 48 h. Until the procedure is routine, especially if a lower temperture is used, the meats should be weighed at 24, 48 and 72 h to determine when they reach constant weight, i.e., when they are dry.
12. Calculate the condition factor from the formula.

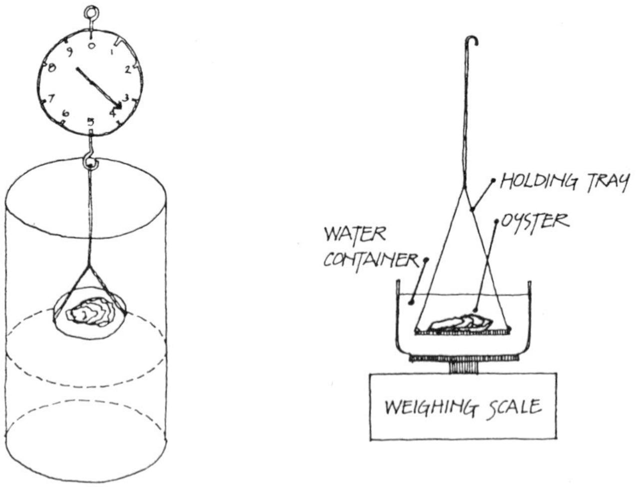

Fig. 18. Apparatus for measuring volume of oysters. Weighing in air and in water (from Quayle, 1980)

Everthing must be weighed in grams, otherwise the volume estimates will be erroneous. The volume of the shell used in the formula above is calculated as the whole volume (#3 - #4) minus the shell volume (#7 - #8).

Condition factor may be measured with a large enough sample of single oysters. For a seasonal study, it is less time consuming if a bulk sample is measured with more individuals. To determine the sample size, it is first necessary to measure the condition factor of a number of single oysters, possibly 25 to 30, to obtain an indication of variability. This variation when applied to the correct statistical formula will provide a rough guide to an adequate sample size for the local situation. The statistical details will not be discussed here. The problem is to determine the changes in average condition. When the population changes and the individual variations are known through the whole season, statistical analysis can be done to determine the appropriate sample size. Until then it is best to just take a large sample, 50 - 100 oysters. When the animals are available it is always safer to use large numbers of individuals to allow for individual variation. The condition factor should be measured on these animals in two to four groups. The agreement of the replicates will indicate whether the sample size is large enough, assuming that your technique is satisfactory.

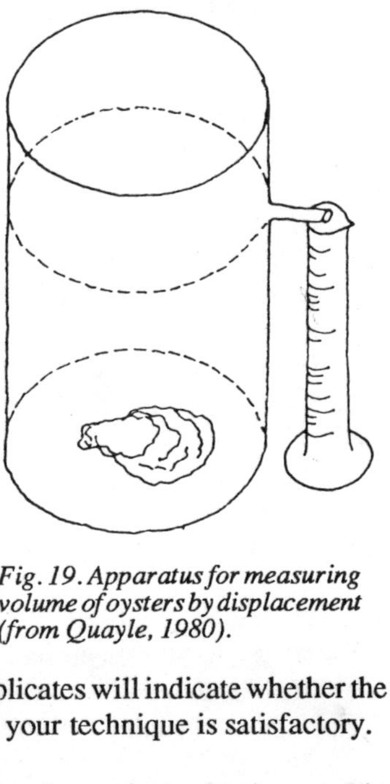

Fig. 19. Apparatus for measuring volume of oysters by displacement (from Quayle, 1980).

This is the standard method of condition factor determination used in most shellfish laboratories and reported in the literature. For local use suitable variations of the method may be used, but comparisons with results from other laboratories may not be possible. For some species there is a fairly good straight line relationship between whole volume and internal volume, so as an approximation the whole volume may be used in the formula, thus eliminating measurement of shell volume. Also the difference between the whole weight and the shell weight in air may be used as an approximation of the internal volume.

As indicated, volumes may be determined by weighing in air and in water. It may also be measured by displacement, measuring the overflow from a vessel filled with water after the oyster or oysters have been placed in it (Fig. 19). This is not as accurate as the previous method but satisfactory if a large sample (10 to 15 oysters) is used. It must be remembered the condition factor, as measured, is only an estimate at best. Weighing and measuring should be done carefully and accurately, but without going to extreme lengths to do so.

In addition to the numerical evidence from condition factor measurements, direct visual observation is also important. The general size of the meat, colour and appearance of the body surface and thickness of the mantle are important criteria and should be noted in data books. The body colour of an oyster in good condition is generally white to cream with the dark digestive gland not visible. If the oyster is in the glycogenous state the surface is smooth, with a thick solid mantle, while if it is in the spawning state the gonadal tubules (vein like in appearance) are quite apparent and the mantle is rather thin. In the post–spawning or spent stage, the body is reduced in size and takes on a somewhat grey colour, with the dark digestive gland readily visible. The mantles are thin and transparent.

If histological methods are used for seasonal gonadal changes, the custom is to divide the process into stages, as follows.

Stage 0. Indifferent. Difficult, if not impossible to determine sex. No follicles present. Mainly Leydig tissue.

Stage 1. Beginning of gametogenesis with the appearance of follicles. Primary sex cells on the follicle walls.

Stage 2. Follicles enlarged and gametes well formed but still not ripe.

Stage 3. Sexually mature with ova capable of fertilization and sperm active.

Stage 4. Recently spawned with follicles partially collapsed with none or only a few relict ova or sperm. This is followed by replacement by Leydig tissue and back to stage 1.

It should be emphasized that these are not distinct phases. Various authors have given slightly different numbers and descriptions to the "stages."

There have been methods described to quantify the rate of gonad development by estimating the percentage of the gonad that consists of gametes.

The development of the gonad in an individual is a continuous process in each follicle. There may not be completely synchronous development of all the follicles within one individual. There will certainly be variation among individual oysters in the stage of development. In some populations there may be very little variation among individuals. However, in most tropical species, one can expect considerable variation among individuals and the usefulness of these indices in monitoring reproduction may be limited. Hence, one should initially use simple methods to monitor reproduction until it is clear that more elaborate methods will provide useful results.

The approach of the breeding season in scallops, usually accompanied by rising water temperatures, is indicated by enlargement of the gonad and an increase in the intensity of colour of the ovary and testes. These characteristics may be observed through the ventral shell gape without destroying the specimen. With several years of observation, a colour chart may be developed showing changes in the gonad colouration as maturation proceeds. The smear and histological methods can be used to confirm the state of gametogenesis. A modification of the condition factor as a gonadal index is a sensitive method for scallops. Since the gonad can be removed easily from scallops, the ratio of the gonad weight to total weight is a usefull index (using dry weights).

Spawning Stimulus

The spawning act has already been described but there is the question of its initiation. One requisite for spawning is temperature, and in temperate waters the gametogenic cycle follows the spring increase in temperature, culminating in ripeness when water temperature reaches the spawning threshold. Also adequate temperatures are required for the development of the larvae. Temperature *per se* is usually not a spawning stimulus but a fairly rapid change may be, as long as it is in a suitable range and the gonads are ripe. Males may spawn at a lower temperature than females. In the temperate zone salinity is relatively stable and does not have the same effect as the more variable temperature.

Artificial induction of spawning is standard practice in laboratories and hatcheries where exact control of temperature or chemical stimulation is possible. Spawning may be induced by the addition or injection of a variety of chemicals, such as ammonium hydroxide, serotonin or chemicals released by the actual gonadal products. The use of serotonin in the hatchery has recently been shown to be more effective that other chemical methods. Gonadal products as stimuli are important not only for oysters but many other marine species as well. The gonadal material taken in during the feeding process stimulates the animals. The ejaculatory nature of the female oyster egg discharge, along with water currents, spread the sex products over a considerable area and leads to the typical mass spawning. It is possible, and it has been done, to artificially induce spawning in an oyster bed a few ha in extent by inducing a number of animals to spawn and spread the gametes over the oyster bed.

The temperature at which oysters spawn varies with the species. *Tiostrea* in Chile and New Zealand may spawn at temperatures as low as 12° C. *Crassostrea* in the temperate zones usually requires 20° C. *Ostrea* may spawn at about 15° C. Those in the tropics seldom spawn below 25° C.

In the tropics, the temperature is a relatively stable factor, while salinity is often highly variable. It is generally considered that salinity, or rather change in salinity, is the spawning stimulus in the tropics, along with the presence of gonadal products. Most tropical bivalves have long spawning seasons compared to temperate species. There is little information on environmental stimuli for gametogenesis and spawning of tropical bivalves. Although changes in salinity are thought to be responsible, it is often difficult to demonstrate a clear relationship as has been done with temperature and spawning in temperate bivalves. In most cases the change in salinity in the tropics is not nearly as regular as the changes in temperature in temperate waters. Careful records of spatfall, temperature, salinity, tidal situation, general weather and other observable phenomena are required to establish relationships with spawning activity and must be done locally for each situation.

Under normal hydrographic and weather conditions, gonadal development generally proceeds steadily although, from year to year, there may be some variations in the time when gonads reach optimum ripeness. Normally, a bivalve well within its geographic range spawns regularly every year, but this is not always the case.

To observe the spawning act, several bivalves are held in a basin of water at the normal breeding temperature and salinity. The gonads of several other oysters are mascerated into a slurry or soup and the slurry is gently forced with a pipet into the inhalant current of oysters in the basin. There is usually a latent period of at least several minutes before spawning begins.

LARVAE

Larval stages

The next stage in the breeding cycle after spawning is fertilization and subsequent development of the egg into a larva. The eggs, when lying in the follicle of the gonad, are pear shaped. When spawned, they assume a spherical shape and in most *Crassostrea* species have a diameter of about 70 µm (0.07 mm). *Ostrea* eggs are slightly larger.

After fertilization, cell division proceeds rapidly, forming within hours the swimming trochophore stage. In about 24 h, probably less in the tropics, two tiny valves have been formed and within 48 h, the shell fully encloses the body, and elementary organ systems have been formed. Soon a definitive swimming organ called the velum is formed. On account of this, bivalve larvae are called veliger larvae (Fig. 20). The velum is a ciliated platform that protrudes outside the open shell and is used as a swimming as well as a food collecting organ. The swimming ability of a larva is limited mainly to maintaining its level in the water column. It is moved about horizontally by water currents. In the first few days the larva has a "D" shape; thus, it is often called a "D" larva or straight hinged larva. Soon protruberances on the straight hinge line develop as the larvae become rounder. These are the umbones and the larva at this stage is called "early umbone." As development continues the umbones of some species become more prominent. At this stage, the mouth, oesophagus, stomach, intestine and digestive gland are well developed, as are the rudimentary gills. In the larval stage all bivalves have two adductor muscles. The larval shell is quite different from the adult

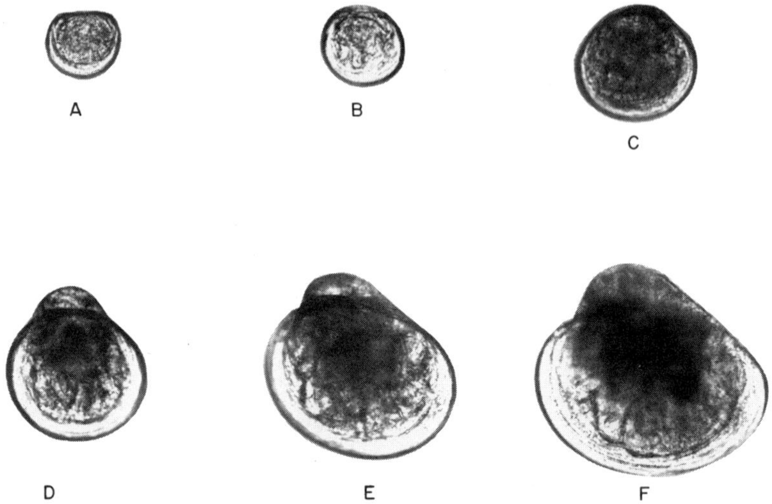

Fig. 20. Stages in the development of the larvae of the Pacific oyster. x117. A = straight-hinged stage (90 μm: B = every early umbone (110 μm): C = early umbone (160 μm): D = umbone (200 μm): E = umbone (250 μm): F = advanced umbone (300 μm) (from Quayle, 1988).

shell. It does not contain as much calcium carbonate and therefore is lower in density. The shell is also almost transparent showing the soft parts of the body.

At this point the oyster larva is somewhat more than half grown and is about 200 μm in length. This is often termed the mid umbone stage. Just before the larva reaches its maximum length of about 300 μm (in most *Crassostrea* larva) an active foot, a cement gland and, on each side of the body, a black eye spot develops, indicating the larva is mature or nearly so (Fig. 21).

The larval period of most species depends on temperature and food supply. In temperate zones for *Crassostrea* it may last from 15 to 30 days and probably not more than 15 days in the tropics (Fig. 22). In *Ostrea* the incubatory period may be about 10 days and the same for the free-swimming time. In contrast, the Chilean and New Zealand *Tiostrea* larvae have a planktonic period of 24 h or less at the end of a long incubation period — the length depending on the temperature.

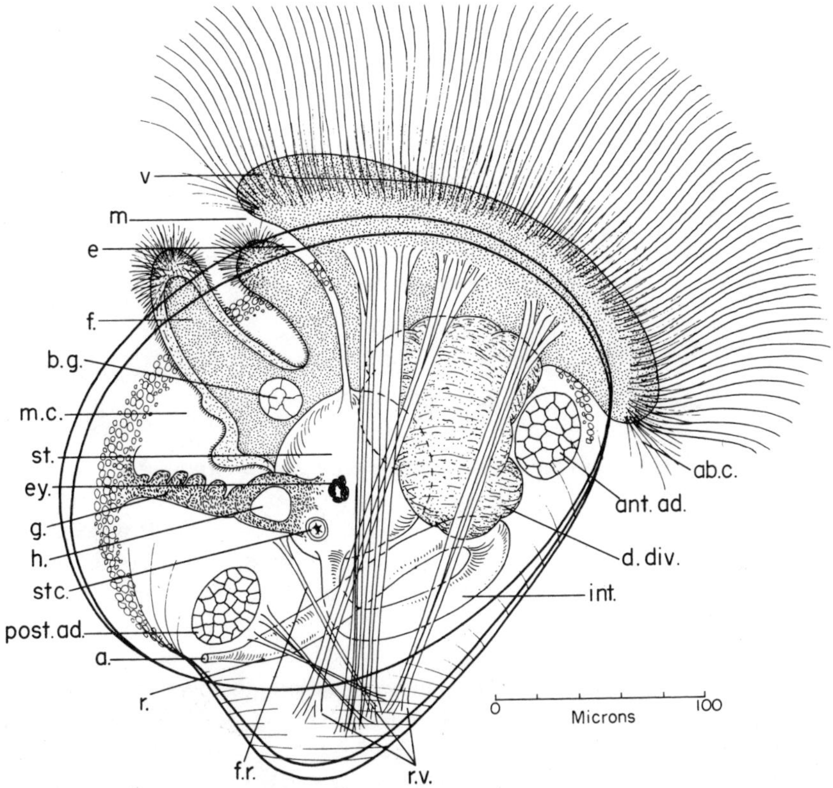

Fig. 21. Optical section of fully developed larvae (pediveliger) of C. virginica *viewed from the left side, in swimming position. Composite drawing from a number of photomicrographs of live, slightly narcotized larvae and whole mounts of glycerol. a. = anus; ab.c. = aboral circle of cilia; ant.ad. = anterior adductor muscle; b.g. = byssus gland; d.div. = digestive diverticula; e. = esophagus; ey. = eye; f. = foot; f.r. = foot retractor muscles; g. = gill rudiment; h. = heart; int. = intestine; m. = mouth; m.c. = mantle cavity; post.ad. = posterior adductor muscle; r.=rectum; r.v. = velar retractor muscles; st. = stomach; stc. = statocysts; v. = velum (from Galtsoff, 1964).*

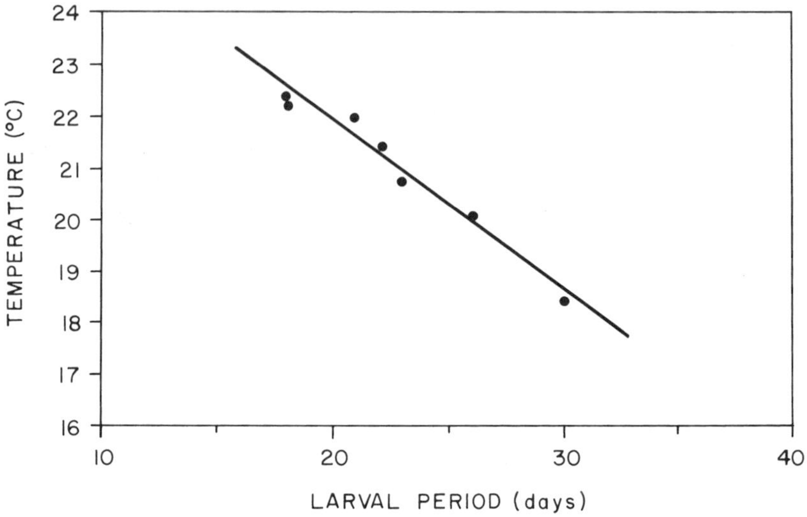

Fig. 22. Larval period of the Pacific oyster at various temperatures. Data from Pendrell Sound, 1950–55 (from Quayle, 1988).

Available food of the right size and quality is important to larvae. This consists of bacteria or other minute organisms, diatoms or flagellates, generally in the order of 10 µm or less. The algae require an adequate supply of nutrients as well as light. There is also increasing evidence of the use by larvae of materials dissolved in the sea water and particulate organic material (detritus).

Mussel larvae, along with oysters, are among the bivalve larvae most readily identifiable. In the straight hinge ("D") stage, the hinge is long in comparison to the height. In the advanced stage, the whole larvae is oval in shape with a blunt, round posterior and a rounded but more pointed anterior. The umbones are nipplelike and small, arising abruptly from the hinge line. The internal hinge shows numerous teeth (about 30), large at the ends and small in the centre, with a ligament at the posterior end of the hinge line. Colour of the shell is quite yellow and the black eye spot is conspicuous. Size of the settlement stage of *M. edulis* is about 300 µm.

All commercial clams are planktotrophic like the oyster genus *Crassostrea* and the larval lives are quite similar, including length of larval life. Size at settlement tends to be under 300 μm, with many as small as 220 μm at settlement.

Spawning and larval development of scallops are similar to those of oysters and clams. The larvae of most scallop species have a characteristic shape and the length at setting varies from 190 to 300 μm, with a larval period of about 2 to 4 weeks. The umbo of advanced stages is not prominent, and at least one species (*Equichlamys bifrons*) has a barely perceptible umbone on a straight hinge line. The general shape is triangular with the anterior end rather pointed as compared to the posterior, but the antero-ventral line tends to be straight rather than rounded. The height of advanced stages approximates the length. There are equal numbers of taxodont teeth at each end of the hinge line. Most but not all pectenid larvae have eye spots.

Larval behaviour

The currents are a major influence on dispersal of bivalve larvae since swimming ability is insufficient to counter the force of currents. This is the mechanism by which larvae are distributed, and studies on introduced

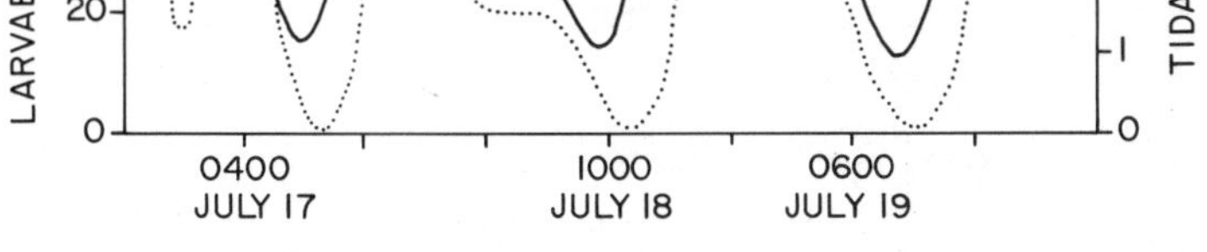

Fig. 23. Relationship between the tidal cycle and percentage of straight-hinged Pacific oyster larvae occuring at the surface. Percentage is the surface fraction of larvae in six samples taken between the surface and a depth of 5.5 m at 3-h intervals. The solid bars along the baseline indicate the periods of darkness. Station 2, Pendrell Sound, July 16-19, 1956 (from Quayle, 1988).

species, such as *C. gigas* in British Columbia have shown that larvae may be transported at least 80 km from the spawning site in one larval period of about 25 days. Despite minimal swimming ability, they are capable of some vertical movement, and a number of studies have shown diurnal activity with movement toward the surface at night and retreat to depth during the day (Fig. 23). This movement may not be greater than 4 to 5 m. Such vertical movement is one explanation why, in most estuaries, the larvae are able to maintain their position in the river instead of being completely swept out to sea. However, as seen in Fig. 23, there is also a correlation with the tidal cycle and it is difficult to separate the two effects. This vertical movement must be taken into account when sampling bivalve larvae.

Particularly in estuaries, the oceanographic situation may be a factor in larval distribution, since haloclines and thermoclines may act as barriers to the vertical movement of larvae. Turbulent or upwelling areas that bring to the surface colder, deeper waters may also produce temperature barriers to the horizontal distribution of larvae.

Larval attachment

When bivalve larvae reach a certain size, about 300 μm in the case of most *Crassostrea* species, and have developed the eye spot and foot, They are ready to metamorphose and become attached. (Note that some species do not develop an eyespot.) The size at metamorphosis varies with the species. This act of attaching to a solid object is called "setting" or "spatting." The term spat is used for the attached bivalve and the word may be used as a singular or plural noun. The collective settlement may be called a "set" or a spatfall. There is no clearly defined time or size when an oyster ceases to be a spat. A spat may also be called a "seed," also used in the singular or plural.

When the oyster larva reaches setting size, the foot and cement gland are fully developed. The larva needs to encounter a solid object such as a properly placed collector in order to complete metamorphosis. If the site is not suitable, the larva may resume swimming. If it is suitable, it may crawl about with its foot and if the surface is clean, not silty or greasy, the larva will force from its cement gland a minute drop, into which it will crawl, always with the left valve in the cement. Within minutes the cement sets and the young oyster, now a seed or spat is fixed for life.

Then comes a dramatic period in the life of the tiny oyster. The foot and cement gland are detached, the velum is lost, the body becomes twisted; the anterior adductor muscle is lost and the posterior one moves more to the centre of the shell. Shell growth at this time is relatively rapid and the new shell, called the dissoconch, is quite different from the larval shell, or prodissoconch.

Mussel settlement, at least in the European *Mytilus edulis*, sometimes takes a somewhat different pattern with the development of a secondary migratory phase in the newly set spat. The term pediveliger is often applied to advanced stage mussel larvae owing to the presence of the foot. (This is not really different from other bivalves, it is just a term used more by mussel biologists!) In *M. edulis* the velum appears to become reduced in the later stages of the larval period and has almost disappeared by the time of initial settlement when the foot, with the byssal gland, is the dominant organ. After initial settlement by byssal attachment (the term plantigrade is often used for this stage), the spat may release this attachment and move to a more permanent site with semi-permanent byssal attachment.

Note that mussels, clams and most scallops are not cemented like oysters to a substrate at settlement. The larva, in its initial settlement stages, tends to select a filamentous substrate such as hydroids, or fibrous material like frayed rope. However, settlement does occur on firm flat surfaces. Hence, oyster spat collectors may be used for experimental mussel spat collection studies. Whether mussels of species other than *M. edulis* undergo the secondary settlement process is not known.

In most clams the byssus is lost soon after settlement but in others it persists in the adult stages, as in *Arca* spp., the ark shells, as well as in a number of other bivalve groups, like some scallops.

Initial settlement of scallops is with byssal attachment and the choice of surface appears nonselective, as long as the substrate is clean. Algae of various species, hydroids and rock are natural setting sites. Fine monofilament synthetics such as fish nets or vexar type mesh are acceptable and are the basic cultches in the commercial collection of scallop spat. Soon after the first addition of dissoconch shell, the byssal notch and ears begin to appear, confirming identification of the spat as a scallop. Byssal attachment may persist until a diameter of 10 mm is reached, when the spat detach and fall to the bottom, if not already there, but this size varies with species. Byssal attachment and detachment continue in some species with decreasing frequency as the scallop grows to maturity, while in others byssal attachment ability is lost after the initial settlement. Size and time of byssal detachment should be determined.

SPATFALL FORCASTING

Spatfall forecasts are required when predictions of the time and intensity of spatfalls are required, either because of infrequent or widely spaced setting or to avoid excessive fouling of collectors. Forecasting in the tropics is sometimes not necessary because of the long breeding season with repeated or continuous spatting.

While not strictly a forecasting method, detection of the earliest spatfall is a positive indication that a spatfall has begun. This method will not indicate the possible magnitude of the spatfall. Nevertheless, it should be used as one of the tools. It requires close surveillance with frequent examination of the exposed cultch.

In temperate waters breeding is synchronized in oysters by the rise in temperature. After a certain period at temperatures above a critical temperature, spawning will occur. Many attempts have been made to mathematically relate breeding with such criteria as temperature-days from a certain date. This may give an approximation of the time of sexual maturity but the occurrence of necessary spawning stimuli or larval survival is difficult to predict from long range. Most spatfall forecasting presently practiced depends almost entirely on plankton sampling. Initially, qualitative sampling in the form of 5–minute surface plankton tows at about 2– or 3–day intervals will determine the presence or absence of larvae. As soon as early stage larvae appear, quantitative sampling is required. This can take the form of metered pump samples or vertical tows if there is adequate water depth. The number of sampling stations and sample depths will depend on the hydrography and topography of the area. A bay or estuary with temperature or salinity stratification requires a different sampling regime from one that does not, as do those with a small tidal range compared to those with a considerable range. Siting of the stations in a bay or estuary studded with islands or headlands will require more sampling stations than one that is topographically simple. Initially, it is preferable to have many sampling stations. With experience, the number can be reduced by selecting those with high significance as indicator stations. A survey of the natural spatfall intensity and distribution may suggest potential sampling sites since this will be an indication, on the average, of how the larvae are distributed.

The frequency of sampling should be high initially, at least once a week. Experience will indicate the feasibility of a lower frequency. This will depend in part on the length of the larval life; the longer the life, the longer the periods between samples. With a brief planktonic period, a shorter interval is needed. As setting approaches, more frequent sampling will increase the accuracy of the forecast.

Actual forecasting is based on the initial number of straight hinged larvae, their growth rate and the rate of decrease of numbers which inevitably occurs, as well as prospective weather conditions. Several years of observation of these factors are necessary before a good forecast can be assured, for it is not possible to transfer experience and numbers from another species or country. As an example, an initial concentration of about 5 straight hinged larva/litre, when reduced to one advanced setting stage larvae/4litres after a larval period of 18 days, produced a spatfall of 1 spat/20 cm^2. However, at another site the number of spat produced would be very different.

Sampling larvae

Oyster larvae are seldom abundant enough to be obtained in significant enough numbers from a simple water sample. To concentrate larvae from a large sample it is necessary to resort to special techniques. This may be done by means of a centrifuge, but the usual method is by sieving. For collecting bivalve larvae a plankton net is used, plankton being floating or weakly swimming plants and animals. The plankton net is a cone of fine meshed silk or nylon cloth with a special weave so the apertures maintain an exact size.

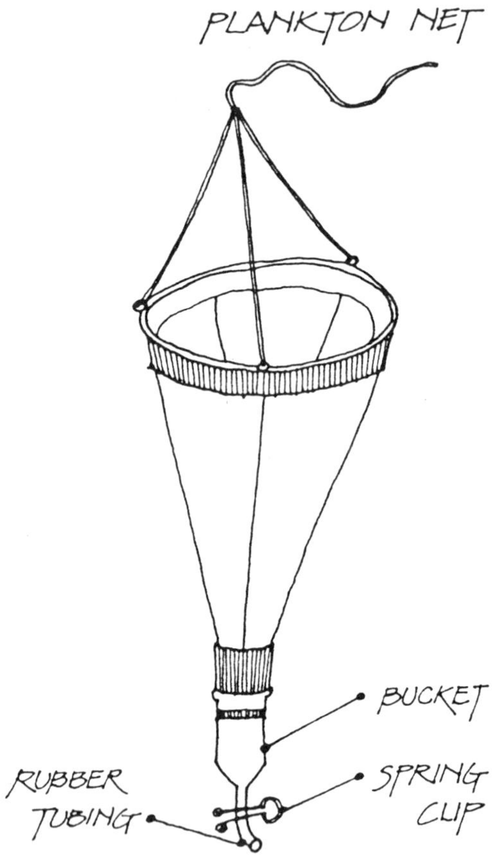

Fig. 24. Plankton net (from Quayle, 1980).

The usual mesh size for bivalve larvae is No. 25 or 20 (64 µm or 76 µm). The mouth of the cone is held open by a metal ring to which is attached a towing or holding bridle. At the small end of the cone is a bucket (in which the plankton is collected) with a tap to allow its contents to be transferred to a jar (Fig. 24).

The simple way of collecting a plankton sample is to tow a net behind a boat whose speed is enough to keep the net just below the surface of the water. The usual tow duration is about 5 minutes. The net is then drawn to the boat and lifted out of the water and allowed to drain. The sides of the net are then washed down with water and the contents of the bucket drained into a collecting bottle into which a label showing the time, date, type of tow, and location has been previously placed. A label on the outside of a bottle may be easily lost. The inside label should be of waterproof paper and written in pencil. Unless for some reason the sample is to be examined alive, the contents are preserved by adding enough concentrated formalin to establish a concentration of 3 to 5 %. After a day or so in the formalin solution the sample should be transferred to a 70 % alcohol solution if it is to be preserved for any length of time, for even buffered formalin will eventually erode the larval shells and prevent identification. If possible, the plankton net should be washed with fresh water after each use.

It is difficult to make an accurate estimate of the amount of water filtered in a short tow. This is because even in a five minute tow the efficiency of the net or the rate at which it can collect can vary greatly between the beginning and end of the tow owing to clogging of the meshes with plankton organisms or silt. Therefore, this type of sample is strictly qualitative and only a measure of presence or absence of a particular larva or whether it is rare or abundant.

For an accurate determination of number of larvae per unit volume of water, it is necessary to know the actual volume of water passing through the net. The most direct method is to pump a known volume through the plankton net. This may be measured with a water meter in the pump line or by filling a container such as on oil drum whose volume may be readily calculated. Any type of pump is suitable. One that is in common use is the 12–volt submersible bilge pump. This can be operated with a portable 12–volt car battery and the pump is lowered to the depth from which the sample is to be taken (Fig. 25). A pumped sample may be taken from several specific depths, which is useful in determining the vertical distribution of the larvae. Sampling at discrete depths such as 0, 1, 3, 6 and 12 m every 3 or 6 h over a 24-h period will do this, although it should be repeated to cover both spring and neap tidal periods.

There are many other types of volumetric plankton sampling devices such as the Clarke-Bumpus net which has an attached flow meter device. The Quayle-Terhune sampler consists of a perforated pipe with a pump and water

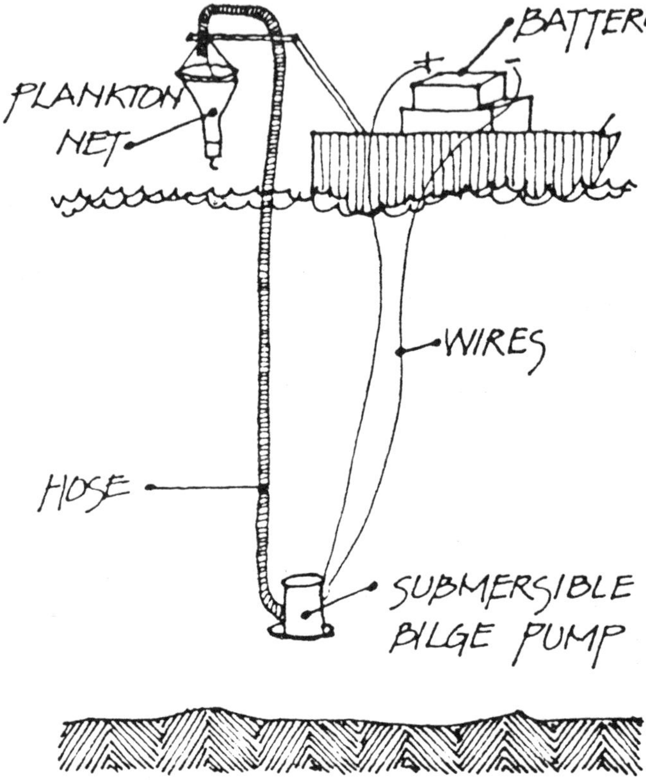

Fig. 25. Plankton sampling with submersible pump (from Quayle, 1980).

meter (Fig. 26). This samples evenly over the length of the pipe, usually 3 m and has the added advantage of also being a moving sampler which, in part, takes into account changes in the horizontal distribution of larvae. Another moving sampler is the Westley device which samples at three discrete depths from a moving boat (Fig. 27).

A simple method of volumetric plankton sampling is the vertical tow. The plankton net, with a weight attached to the bucket, is lowered to the required depth and then lifted to the surface. Care must be taken not to move the net too rapidly, so all of the water is allowed to flow through the meshes of the net. If moved too rapidly filtration will not be complete. This type of

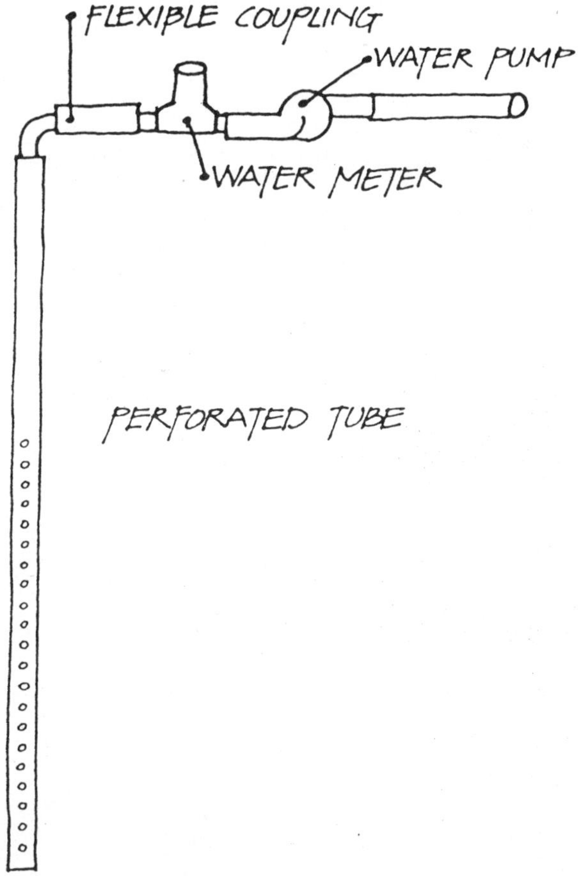

Fig. 26. Quayle - Terhune plankton sampling device (from Quayle, 1980).

sample takes into account variations in the vertical distribution of the larvae and a water volume measuring device is not necessary. Knowing the area of the mouth of the net and the depth through which it is drawn, the volume may be calculated as follows:

Volume = $\pi \times r^2 \times d$

where r is the radius of the net, d is the depth and $\pi = 22/7$.

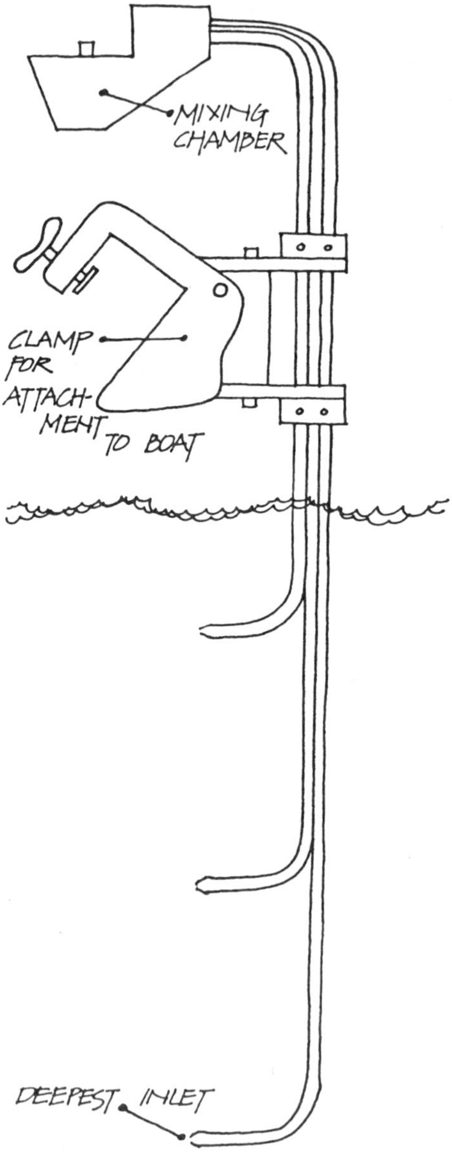

Fig. 27. Westley plankton sampling device (from Quayle, 1980).

Example: A net 14 cm in mouth diameter has an area of (22/7) x 72 = 154 cm^2. If the depth of the haul is 10 m (1000 cm), then the volume filtered will be 154 x 1000 or 154,000 cm^3 (154 litres).

If the depth of the area being sampled is greater than about 5 m, a vertical tow is the recommended method of volumetric sampling since it is uncomplicated and rapid as well as requiring little equipment. Care should be taken to avoid having the net drag on the bottom and be damaged.

In initial stages of research, sampling should be restricted to simple, qualitative methods until species identity is certain. Simple methods will give a relative measure of seasonal abundance. This can be used to evaluate the need for costly and time–consuming quantitative measures of plankton sampling. It will also be necessary to do surveys to determine the geographic distribution of larvae before effort is expended on quantitative methods.

Counting larvae

To examine plankton qualitatively for bivalve larvae, several pipet samples are taken from the bottom of the jar of plankton and placed in a 10 cm diameter watch glass. The contents are then swirled to the centre of the watch glass by rotating it in a small circle. The watch glass is then tilted and the floating material moves to the edge where it can be removed with a pipet, leaving the heavier larvae in the centre and on the bottom. Additional clean water may be added and the process repeated until only the heavier material is left lying on the bottom of the glass in clear water.

The sample may be examined with a stereoscope microscope at a magnification of about 50x, adequate for the identification of most larvae. A compound microscope is not needed unless there is some special factor such as hinge teeth to be studied.

For counting larvae a special dish is useful (Fig. 28). These are not manufactured so it is necessary to construct them locally with 5–mm thick plexiglass (acrylic). It is simply a base about 10 x 15 cm with a rim 15 mm high glued to it with acetone, an acrylic solvent. The interior of the dish is divided by plexiglass strips 10 mm high attached alternately to the ends of the dish so there is a continuous lane about 10 mm wide from one end of the cell to the other. The actual distance between the lanes is adjusted so the field of view at the desired magnification under the stereomicroscope covers the lane. Runners are added to the bottom of the dish to prevent scratching.

Accurate subsampling of jars is difficult, so, if possible, the whole contents are put into the counting chamber. In the sampling jar the larvae and the heavier particles lie on the bottom after settling, so much of the supernatant fluid can be decanted and discarded. The remainder is swirled and then poured evenly into the counting cell. A small amount of water is added to the jar, swirled again and poured rapidly into the counting cell. This is done to remove as many larvae as possible from the walls of the jar. Counting begins at one corner of the cell and the lanes are followed to the other corner. If the number of larvae is large, alternate lanes may be counted. Sub-sampling may also be done from the plankton jar which can be filled to a known standard volume and a specific fraction removed either by decanting or with a volumetric pipet after the contents of the jar are well stirred. This must be done carefully and very quickly as the older larvae settle rapidly.

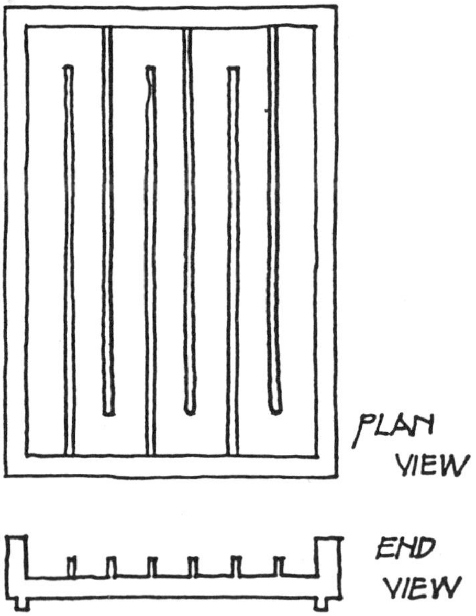

Fig. 28. Plankton counting cell (from Quayle, 1988).

If the whole sample is counted, the total number of larvae is divided by the volume of water filtered through the net. For example, with a count of 3000 larvae and the volume is 300 litre, then 3000/300 gives 10 larvae/litre.

If subsampling is needed the plankton sample is made up to a specific volume in the sample jar, i. e., 400 ml. If a subsample of 25 ml is taken and 800 larvae counted, then there would be 400/25 x 800 = 12,800 larvae in the volume filtered through the net, for example 154 litres. This gives a count of 12,800 divided by 154 or 83.1 larvae/litre.

Larval identification

Identification of bivalve larvae is one of the more difficult aspects of shellfish biology to master. Even when a larva is accurately identified, separating it from the numerous other species of similar shape and size, particularly in the early stages, presents a problem. A general, ongoing study of bivalve larvae, in addition to the species of concern, is necessary to make identification and counting more than a guess.

In some instances larval study may not be necessary but for a culture to achieve its optimum potential some knowledge is required. It may be possible to conduct a satisfactory culture without spatfall prediction, but when fouling or cultching timing is critical, it may mean the difference between adequate and inadequate settlements.

There are several ways to identify molluscan larvae. The most positive is the culture of the larvae in the hatchery or laboratory. If the facilities to culture larvae are not available it may be contracted out to another laboratory or hatchery, even in another country by supplying adult animals, preferably in the spawning or near spawning state. In this way illustrations and specimens of all the larval stages and spat may be acquired.

A second method is useful in the tropics where air and water temperatures may not be too different. A number of advanced stage larvae suspected from published illustrations to be the species of concern, may be isolated from live plankton samples and placed in a 10- to 15- litre container of filtered seawater using the finest plankton netting available (64 µm or less). This removes the larger plankton organisms but retains the larval food. The water should be stirred or aerated and replaced every other day by filtering through a plankton net to retain the larvae. There is usually enough food in the filtered water to sustain the larvae until metamorphosis and spatting.

The third method consists of matching advanced stage larvae from the plankton with the larval shell (prodissoconch) of the smallest spat that can be collected. Size and shape comparisons may be made. The relative abundance of a larva in the plankton and of a species undergoing spatting is also an indicator. A knowledge of the local molluscan fauna and the relative abundance of species is useful in this connection. In this approach, one works "backwards" by first identifying the spat through growth studies where the various spat forms are cultured to a size that can be positively identified. Then the connection is made between the larval forms and the spat. The literature now contains good illustrations of many bivalve larvae that will be useful as guides. The publications of Loosanoff, Davis and Chanley (1966), Chanley and Andrews (1971), Rees (1950) and Lutz et al. (1982) are particularly useful. Rees' paper is useful for placement of a larva in its family.

Bivalve larval identification is based mainly on shape but also on characteristics such as colour, unusual characteristics such as the byssal notch in some anomiids, or pigment or eye spots, most often found in anisomyarian species such as oysters and mussels. Drawings and sketches of larvae, always with measurements attached, are useful, including some with some outstanding characteristics slightly exaggerated. Photographs provide the most accurate depiction of the larval shape for the thickness of a line in a drawing can alter the appearance. However, in a photograph how the larvae happens to be tilted will alter the appearance. This problem can and should be avoided by photographing a single valve with the rim downwards. Single valves may be obtained by shaking the larvae in a 1% solution of potassium hydroxide or in a bleach solution such as Perfex or Chlorox. The chemical clears the larvae leaving only the shell visible. These valves with the rims down will show the true outline while those with the rim up will show the provinculum and hinge teeth which are also diagnostic features (Figs. 29 and 30).

Another diagnostic feature involving size is the division of the larval shell into two sections. The early shell of the "D" shaped stage is termed Prodissoconch I and is marked off by a fairly distinct line from the older shell, termed Prodissoconch II. The adult calcareous shell, formed after settlement, is termed the dissoconch. The final length and height of Prod I and Prod

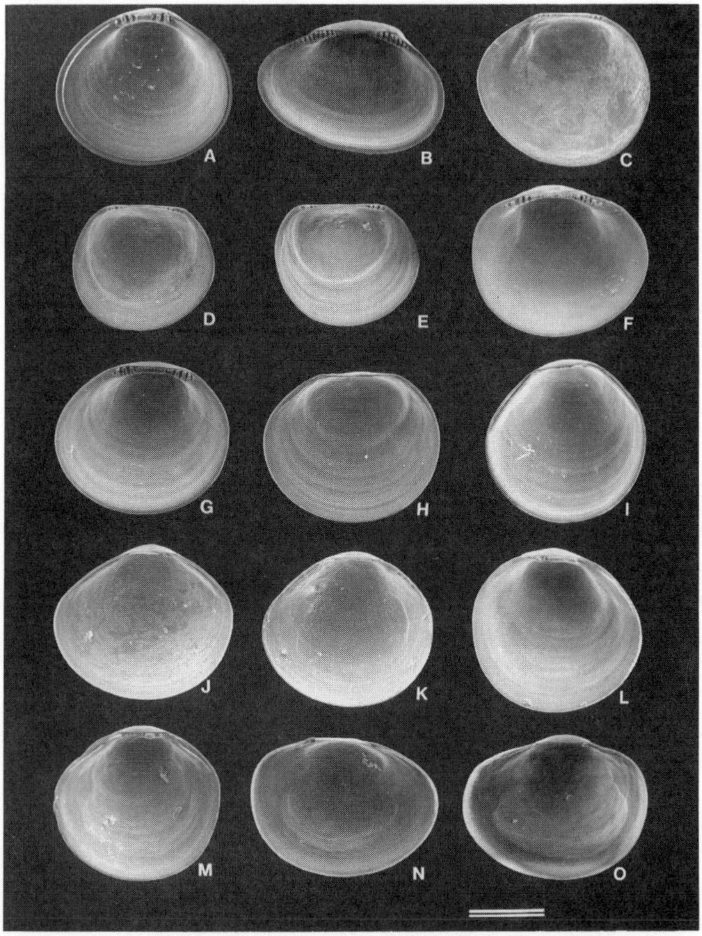

Fig. 29. Scanning electron micrographs of disarticulated shell valves of planktonic larvae of various species of bivalve molluscs. A. Crassostrea virginica *(right valve; mature larva). B.* Arca noae *(right valve; mature larva). C'.* Argopecten irradians *(right valve; mature larva). C.* Argopecten irradians *(left valve; straight-hinge larva). D.* Placopecten magellanicus *(left valve; straight-hinge larva). E.* Mytilus califirnianus *(left valve; mature larva). F.* Geukensia demissa *(right valve; mature larva). G.* Arctica islandica *(right valve; mature larva). H.* Mercenaria mercenaria *(right valve; mature larva). I.* Mya arenaria *(right valve; mature larva). J.* Mulinia lateralis *(right valve; mature larva). K.* Spisula solidissima *(left valve; mature larva). L.* Spisula solidissima *(right valve; mature larva). M.* Ensis directus *(left valve; mature larva). N.* Ensis directus *(right valve; mature larva) (from Lutz et al., 1982).*

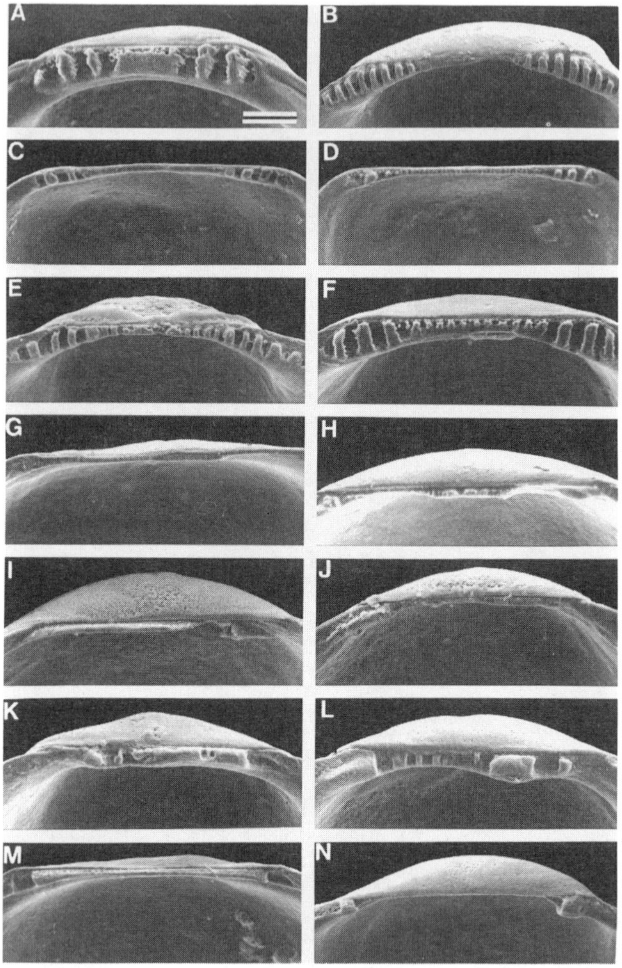

Fig. 30. Scanning electron micrographs of the hinge region of the disarticulated shell valves of the specimens depicted in Fig. 29. A. Crassostrea virginica *(right valve). B.* Arca noae *(right valve). C.* Argopecten irradians *(left valve; straight-hinge). D.* Placopecten magellanicus *(left valve; straight-hinge). E.* Mytilus califirnianus *(left valve). F.* Geukensia demissa *(right valve). G.* Arctica islandica *(right valve). H.* Mercenaria mercenaria *(right valve). I.* Mya arenaria *(right valve). J.* Mulinia lateralis *(right valve). K.* Spisula solidissima *(left valve). L.* Spisula solidissima *(right valve). M.* Ensis directus *(left valve). N.* Ensis directus *(right valve). Scale bar (= 20µm) in A applicable to both Figs. 29 and 30 (from Lutz et al., 1982).*

II are diagnostic for each bivalve species. Prod II usually has distinct growth lines while Prod I does not.

The hinge area of the larvae, thickened to form a base for the hinge teeth and ligament, is called the provinculum. The Solenacea (razor clams) is the only group that contains some species whose larvae have an external ligament. All other species have internal ligaments. The Lucinacea and Erinacea, usually with quite large larvae and often larviparous, have no hinge teeth as is the case with the Chilean and New Zealand oysters (*Tiostrea*). Oyster larvae are quite distinctive with prominent umbones. *Crassostrea* larvae have prominent umbones that are opisthogyrate, being twisted posterior to the centre line of the hinge. *Ostrea* larvae have large broad umbones that are orthogyrate (Fig. 31), being centrally placed on the hinge line. By contrast, *Tiostrea* has no umbones.

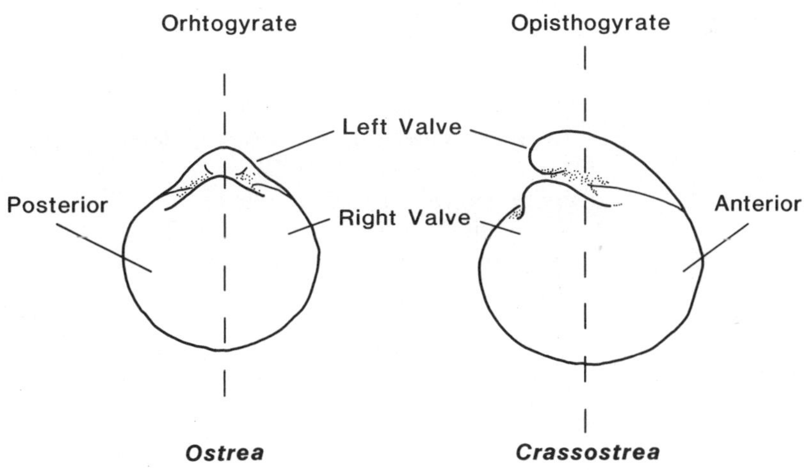

Fig. 31. Umbonal position of the genera Ostrea *and* Crassostrea.

In summary, bivalve larval identification is based upon:

1. Length.
2. Height to length ratio.
3. Shape of anterior and posterior ends — rounded or pointed.
4. Position and shape of the umbones — opisthogyrate or orthogyrate and prominence.
5. Colour. This pertains to certain larvae only, for some are colourless or variable depending on food. In general, live larvae of the genus *Crassostrea* are brownish while *Ostrea* larvae are black. Some larvae of the Veneridae have yellow valves.
6. Length and shape of Prod I.
7. Teeth number and arrangement.
8. Ligament shape and position. The typical shape of the larvae of a number of molluscan families is shown in (Fig. 29).
9. Special characteristics such as the notch in some Anomiidae.

Length of larval life

This is most easily determined by laboratory or hatchery culture but may not necessarily represent the situation in nature. In the same way the length of the larval life in 1 year might be quite different from that in another or at different times within a breeding season. Since temperature conditions are relatively stable in the tropics, the variation may not be so great as in the temperate zone. Temperature is the dominant factor in determining length of larval life and may cause a difference of as much as 10 days.

In cases where individual spawnings are separated in time it is possible to follow a single brood in the plankton from the straight hinge stage to spatting, thus giving the length of larval life. By measuring a sample of 100 larvae each day through the larval period it is possible to construct a growth curve (Fig. 32).

If spawnings are not distinct but vary in time it may be necessary to measure daily samples and develop frequency polygons. The mode on these may be followed until it reaches the setting size for the species (Fig. 33).

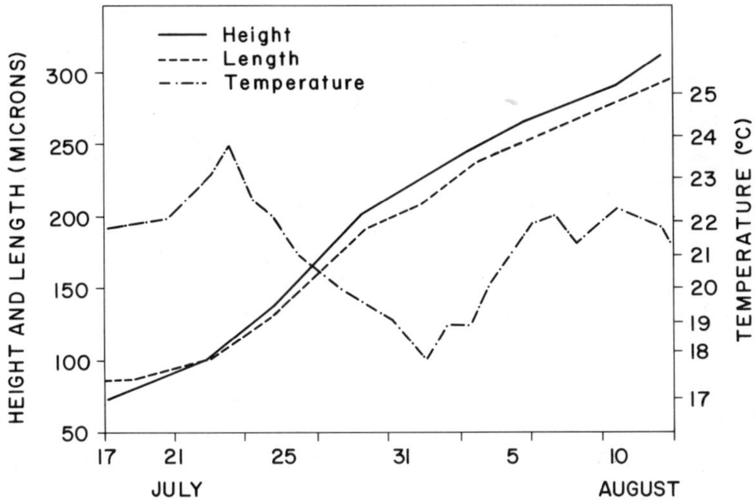

Fig. 32. Growth rate of larvae of Pacific oyster in Pendrell Sound, Aug. 1955 (from Quayle, 1988).

SETTING BEHAVIOUR

Some knowledge of the setting behaviour of larvae is necessary for successful spat collection. Among the factors that may influence setting behaviour are:

1. Temperature
2. Salinity
3. Light
4. Tidal phase
5. Lunar effects
6. Depth
7. Angle of cultch surface
8. Cultch texture
9. Colour of cultch
10. Cleanliness
11. Current speed
12. Gregariousness

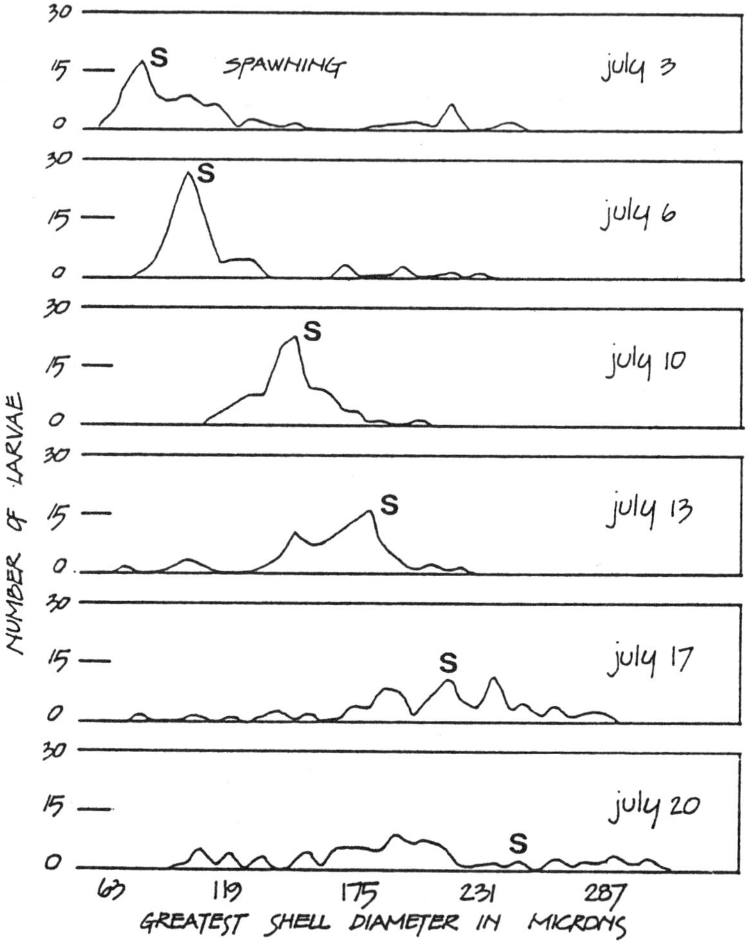

Fig. 33. Length–frequency graphs to determine length of larval period (from Quayle, 1980).

Temperature and salinity

These are not significant factors under normal conditions but may become so in the case of a wind storm causing upwelling of deep cold water or heavy rain or floods resulting in drastic salinity changes beyond the tolerance of the larvae. The presence of thermoclines or haloclines will also affect the vertical distribution of larvae and therefore the positioning of collectors.

Light

Light and its effects vary according to the time of day, turbidity of the water and cloud cover. The major effect of light on bivalve larvae that has been studied is diurnal movement, with a tendency to move near the surface during the night and deeper during the day (Fig. 23). The main point is for the cultch to encompass the depth range covered by the setting larvae. If for some reason setting takes place at a particular time of day or night, the cultch should be placed at the depth where the larvae are expected to be. This may be investigated by exposing cultch at brief intervals such as 3 h, possibly timed to coincide with time of slack water and midway between. The study should also be carried out during both spring and neap tidal periods.

The effect of light on the orientation of cultch surfaces is somewhat controversial. The general impression is that the underside of horizontal surfaces collect better than upper surfaces but this is not always borne out by experimentation. A higher level of silting on the upper surface of commercial cultch inhibits settlement there. It is difficult to separate the effect of light from the effect of siltation. In practice, however, the type of cultch available, its packaging and facility in handling take precedence over detailed positioning of the cultch surface.

Tidal Phase

While there is some indication that the phase of the tide affects settlement, it is necessary to accept whatever the results of these may be. Cultch will be exposed for many days and, thus, many tidal cycles.

Lunar effects

Phases of the moon affect the breeding of several invertebrate animals but has not been demonstrated for bivalves. There may be indirect effects through the lunar relationship with tides. This variation will become apparent during detailed monitoring of spatfall periodicity.

Depth

Adult bivalves are usually found either intertidally or subtidally, depending on the species. There is often zonation within the intertidal zone. Thus, depth is an important factor both intertidally and subtidally for the settlement of larvae. It may be assumed that settlement will take place in the intertidal zone at all levels, from the highest to the lowest, wherever suitable settlement sites occur. However, not all of these sites are suitable for survival of the spat. This may be determined in part by observation of the natural spatfall, as on the mangrove roots, but also by placing experimental collectors at various tidal levels to examine both settlement and survival. The main factor inhibiting settlement is silt and in the intertidal zone there is typically a gradient from high silting at lower levels to low silting at higher levels of the zone.

The effect of depth on collectors suspended from rafts or longlines will be influenced by the presence or absence of either temperature or salinity stratification of water. However, the depth of optimum settlement is readily determined by suspension of long strings of collectors and counts of the spatfall on the collectors in each 30 cm of the string (Fig. 34).

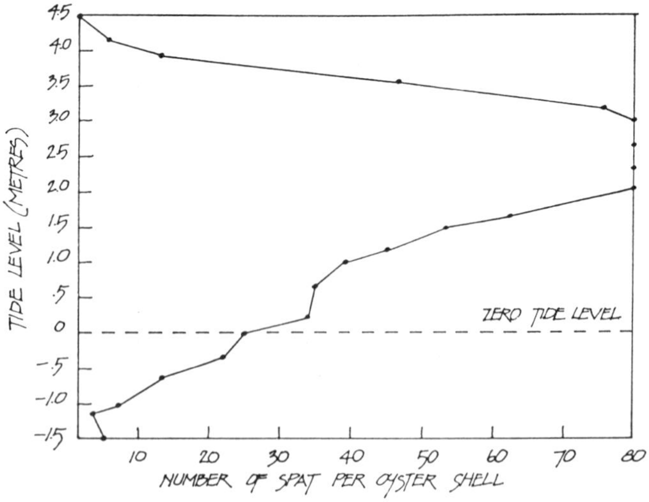

Fig. 34. Vertical distribution of spatfall on fixed cultch (from Quayle, 1980).

Angle of Cultch

As indicated previously, there is the impression that under surfaces of cultch collect better than upper ones. Specific experiments to determine the effect of angle place collector at specific angles around a central axis with the following designation have been carried out.

0 ° — lower horizontal
45 ° — lower surface of a 45 ° surface
90 ° — vertical
135 ° — upper surface of a 45 ° surface
180 ° — upper horizontal

The results of such a study for *Crassostrea gigas* is shown in Fig. 35 and are typical of a number of such experiments. As can be seen, there is little difference between the upper and lower horizontal surfaces. In this instance silting and fouling were not factors because of the short duration of the experiment and absence of turbidity. Here again the practicalities of seed collection, e.g., irregular surfaces, methods of suspending cultch, etc.,

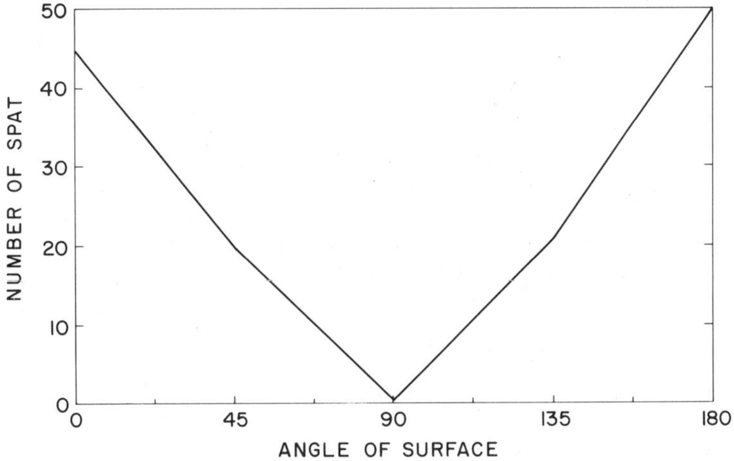

Fig. 35. Relationship between the angle of setting surface and spatfall. Pendrell Sound. 0 °= lower horizontal; 180 °= upper horizontal; 45 °= lower diagonal; 135 ° = upper diagonal; 90 ° = vertical (from Quayle, 1988).

generally supercede specific attempts to angle the cultch and selection of upper or lower surfaces may vary from year to year.

Cultch texture

Oyster larvae require a hard, clean, non-greasy surface and will set on smooth surfaces such as glass or plastic. However, roughened surfaces appear to be more suitable and the spat will frequently be found in cracks or crevices if these are available. This may be one reason cement surfaces are very good. On the other hand, oyster shell is probably the most used cultch in the world and it is quite smooth although spat are frequently found in crevices such as those on the adductor muscle scar.

Colour

Many experiments have shown that colour of the collector has no significant influence on the setting behaviour of oyster larvae. In specific instances the colour may effect survival by effecting the amount of sunlight absorbed (in intertidal cultch) and thus the temperature of the collector.

Cleanliness

Knowing the basic requirement for oyster larvae to have clean hard surfaces for solid attachment, it is not difficult to realize that cultch whose acceptable surface area is reduced by mud, sand or other materials is not efficient. Clean cultch is particularly important when the setting intensity is low. There is a supposition that for cultch to be acceptable to larvae, a bacterial film is necessary but experience in the field and with the eyed larvae setting process indicates that this is not essential, at least for the genus *Crassostrea.*

Current Speed

It is known that high current speeds inhibit settlement but this difficulty is alleviated by slack water and neap tide periods. The hydrography of most areas is complex. In some cases an area of little current flow may be a "trap" for larvae. Each case has to be studied.

Gregariousness

There appears to be a tendency for larvae to settle more readily on surfaces on which there is already some spat. Certainly there is much evidence for a "clumping" phenomenon where oysters occur in groups on the shore rather than in even distribution. Cultch on which oysters have been previously attached appear more attractive to larvae than those with no previous spat.

CULTCH

In nature, oysters tend to set on many different substrates. Usually they are hard surfaces like other oyster shells, rocks, sticks, etc. However, when such hard substrates are not available, the larvae will even set on seagrass. Given a choice, some species of oysters prefer a calcareous substrate like shell.

The material used to collect spat is called "cultch" or "collector." The cultch must be a clean, hard material. The cultch (collector) most used in the oyster industry is the empty oyster shell. It is usually available and it meets several of the major attributes of a cultch — fairly cheap and acceptable to the larvae. Many other materials such as rope, bamboo, rubber tires, limed tiles, cement or tar coated objects, as egg trays, wood veneer, laths, coconut shells and many others too numerous to mention may also be used as cultch.

It is necessary to adapt the cultch to the culture system. This is easier than adapting the culture system to the cultch. Suspended culture requires a larger and more durable individual cultch piece than bottom culture. For bottom culture, the smaller the cultch piece the better (commensurate with the ability of the ground to hold it). Large pieces of cultch are less apt to sink into soft bottoms than small pieces, but oyster clusters from small pieces are more readily broken up to produce single oysters. For bottom culture less durable cultch is better and a disintegrating cultch is ideal.

Some important characteristics of an ideal cultch are:

1. Low cost and availablity in large quantities.
2. Solid, slightly rough, clean surface (the color is immaterial).
3. Light in weight for handling but with a specific gravity high enough to sink.
4. Easily and cheaply packaged for transportation and for suspended or beach cultching.
5. Maximum surface area per unit of volume.
6. A free flow of water to all surfaces with pore diameter adequate for growth of the spat to a size at which it may be separated from the cultch.
7. Preferably adapted to both suspended and bottom culture or to whatever culture method is used.
8. Limited silt collection propensity.
9. For bottom culture a disintegrating cultch is usually preferable.

Spat collection is an essential stage in bivalve culture. However, it is a short period in a long process. Thus, the cost of materials and the labour involved in spat collection are critical. Large volumes of expendable material must be available. Dependence on imports from other countries or regions is not advisable. Low cost, locally available cultch is essential. However, the material must suit the culture system. Thus, it must be attractive to the larvae (solid and clean) and present a large surface area without encouraging heavy siltation. It must be of a nature that makes it easily handled in the following growout stages.

An important factor is the shape of the cultch. Sometimes this may be adjusted as in stick culture but may not be possible at other times as in the case of shell cultch. The reason for concern about shape is the problem of xenomorphism in oysters, which is the propensity of the lower valve to assume the shape of the material on which it is growing. If the oyster is allowed to grow uninterruptedly on a flat surface, the lower valve will not cup and this reduces the space within the shell for meat production. Thus, a general rule is to keep the cultch as narrow as possible. If the spatfall is always dense the narrow cultch is not as important,1 for competition for space will force the oysters up on the umbones as they grow and give greater freedom for cupping. Oysters flat on cultch surface are also difficult to remove if this is part of the culture process. The use of a narrow cultch is well exemplified in the Australian and New Zealand system of stick culture.

The search for the ideal cultch is a perennial one. Most often attention is paid to the use of naturally occurring materials such as shell or wood such as mangrove branches or bamboo. A number of attempts have been made to modify or manufacture cultches. The lime covered roofing tiles once so common in Europe is an instance of modification, as is the cement coating of cardboard egg separators or cement or tar-pitch coated coconut shells. In France specially manufactured plastic forms are coated with cement. Another modification is the cement–coated wood veneer collars. Grooved plastic pipes are specially manufactured for stick culture, whether on racks or suspended. One attempt has been made to manufacture a disintegratable cultch on basic principles from cement, asbestos, calcium and stearates formed under high pressure. The shape is curved like an oyster shell and the required decomposition time can be varied according to the need by changing the composition. This cultch works well in practice but economics have prevented its industrial use.

Bamboo has been used extensively in the past in Japan and is an excellent spat collector as long as it is absolutely dry before use. It is an obvious cultch for the tropics either as sticks or small blocks on strings. Its main objection for strings is its low density thus requiring a weight at the end of the string, but it is ideal for stick culture as it has been used in Japan and recently in the Philippines.

Packaging Cultch

Oysters, scallop or clam shells were originally packaged in the form of strings. A hole is punched in the centre of each valve and these are strung either on a line such as nylon or on wire. Hole punching was done mainly by hand but there are semi-automatic shell punching machines. Shells could also be packaged in trays with a wooden frame and a mesh top and bottom. Shell strings are usually made up in 2-m lengths with 100 to 200 valves depending on their size. The string may be of tightly packed shells or shells separated by spacers made of short pieces of bamboo or hose. The tightly packed shells will have to be separated for growing out (either on strings or on bottom). Strings with spacers may be grown out without further handling. Often the strings are made double with a blank space in the middle so the strings may be draped rather than tied or hung.

Shell bags have been made of used fish nets. With the advent of relatively inexpensive plastic mesh bagging provided in endless rolls, shell strings have virtually disappeared in high labour cost areas. The main difficulty with this sort of packaging is that the lack of a hole in the shell does not make it satisfactory for suspended culture. This is overcome by inserting the spatted shells in the lay of double lay thin nylon rope, a method originally used in Japan with straw rope before the use of galvanized wire. The main caution about bagging shell is that the diameter of the bag must be small enough to allow water flow and, thus, penetration of the larvae to the centre. This diameter will depend on the size of the cultch. The smaller the cultch the smaller the bag diameter should be.

Bamboo for cultch (absolutely and completely dry, otherwise it becomes slimy and unattractive) is probably best exposed in stick form. If it is to be used for string culture it may be cut into appropriate sized pieces. Sticks are usually packed in bundles, with each separated by a spacer (Fig. 36). Here again the spacing of individual sticks and layers must be far enough apart to allow larval penetration.

Other types of cultch (plastic collectors, coconut shells, pieces of rubber tire, etc.) are strung or stacked as seems appropriate. The cemented veneer collars can be simply strung on wire or rope. Heavy cultch like tiles are stacked on the bottom or on racks.

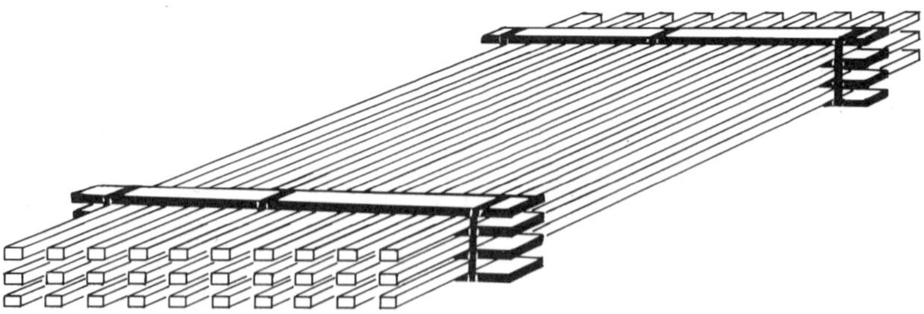

Fig. 36. Bundle of stick cultch. Note that the layers of sticks are separated by spacers and bound together at the ends and in the middle of the spacers.

A consideration with cultch types is the transportability both to the spat collection site from the preparation area and from the spat collection site to the growing site, for these two are not always together. Shell strings tangle easily so they must be carefully piled when transporting. Bags are more easily transported. The weight is a factor in the cost of transporting the cultch.

Exposing Cultch

Cultch may be exposed by placing it on the bottom, or on racks as well as suspending it from rafts or long lines. The timing of exposure depends on the breeding habit of the species. It may be a long season with spawnings either evenly distributed or occurring in several peaks or a short breeding season with one or two peaks. The tropical situation is usually the former and the question of exposure time is often not critical relative to the collection of a spatfall adequate for culture purposes. The problems here relate to other aspects of the culture sequence. Time of spat collection may be related to harvest time (or times) which is related to when meat production is optimal (high condition factor). The fouling sequence is important in relation to oyster spatfall and seed collection. These factors must be assessed in each situation. These problems will be discussed later.

Once the spawning and the setting pattern are established, spatfall forecasting may not be necessary. If a spatfall is missed, there is usually another fairly soon. This will have to be determined empirically in each situation.

The spacing of collectors must allow adequate water movement among them to allow larval access to all the cultch. For intertidal bottom placement the substrate must be firm to prevent sinking and silting. It is preferable to place bags or shell strings in a row, low in the intertidal zone with at least one metre between the rows. For subtidal bottom collection the area should be marked off with stakes driven into the bottom and visible at high tide or with buoys. This enables collectors to be dropped off at intervals along definite lines.

For suspended exposure, cultch units should be separated by a distance of about four times the diameter of the collector if the area has a moderate current flow. If there is virtually no current, spacing should be somewhat greater. Experience and testing will be necessary for spacing depends on setting intensity as well as water flow. Collectors in the centre of rafts usually do not collect as well as at the perimeters. The long lines is a superior system for spat collection as there is usually adequate spacing between lines.

Placement of racks to collect spat is largely dependent on tidal range, setting behaviour of the larvae and amount and location of fouling. Natural spatfalls on mangrove roots or rocky shores provide an indication of setting levels, but this is often the result of availability of a setting surface rather than tidal level or site. On mangrove roots relatively high in the tidal zone, what is seen is to a high degree the survival of the hardiest spat. Larvae settle at the highest tidal level if there is a suitable cultch but do not survive. Usually fouling is least in the higher tidal zone levels, except possibly for some species of barnacles. Normally, placing cultch at the lower tidal levels is most advantageous for survival, growth rate and spatfall intensity. In temperate waters it has been shown that floated cultch will collect more spat than intertidal cultch but has the disadvantage of the possibility of more fouling. It is also more costly to float cultch. Thus, there is a need for more precise timing of cultch exposure.

The timing of removal of settled cultch from seed collecting to growing areas is of some importance, particularly with floating cultch. This is not so for shore cultch which has already acclimated to air exposure. Floated cultch should be moved before the spat has reached a size where they impinge on the adjacent pieces of cultch in a shell string or bag. Otherwise all of the cultch pieces will be attached to each other. Moving should be done during cool and damp periods. If it is to be planted intertidally, the moving should be done during a neap tidal period when they will not suffer long periods of exposure during the extended low tides of a spring series. During transportation the seed should not be held in water but rather just covered with sacking or a tarpaulin and kept damp. Aside from the fact that shipping sufficient water to submerge the seed would be very difficult, it is also dangerous to ship bivalves (seed or adults) submerged. The animals soon use up the dissolved oxygen and die. When shipped out of water they may "air breath" by opening the valves. However, they must be kept damp to avoid dessication.

The classic example of shipping oyster seed is the shipment of Japanese oyster seed (*Crassostrea gigas*) to the west coast of North America. In Japan the seed was collected on shell cultch during the summer. During the winter it was held on intertidal racks, the so called "conditioning and hardening" period, acclimating the seed to long periods out of water. During this time growth of the spat (between 3 and 10 mm in diameter) ceases and the shell tends to thicken. Hardening is done for all spat whether for local use or export. In February and March while temperatures are still cool the seed was removed from the strings, sorted and packed into wooden cases (90 x 45 x 30 cm) containing 121.5 litres (4.5 ft^3) with 12,000 to 15,000 spat. These were stored on intertidal racks until enough had been accumulated to make up a shipment of 5 to 10 thousand cases. The seed was then loaded on freighters as deck cargo covered with rice matting, kept cool and damp, but not wet, by occasional sprinkling. The ocean voyage across the Pacific required about 12 days, and up to 40 days may have elapsed between time of packing and final planting of the seed on oyster beds. Thus, shipment of seed over short distances should pose no significant problems. Up to 100,000 cases of have been shipped across the Pacific in the past. With the advent of local supplies of natural seed on the west coast of North America and development of hatcheries, Japanese seed importation has ceased.

HATCHERY SEED

Significant interest in hatcheries as an alternative to natural supplies of oyster seed began in the 1940s. The techniques have now become quite standardized, although success can vary somewhat from hatchery to hatchery and from time to time in the same hatchery. Hatchery construction and operation have been well documented. The most essential element is a water supply of the best quality, though most hatcheries purify the water by filtration and ultraviolet treatment. If the oysters are not ripe they may be conditioned by holding them at elevated temperatures if necessary and by feeding. If this conditioning period is long, egg quality and, thus, larval viability suffers. Spawning is usually not a problem with stimulation either by temperature manipulation or chemicals. A critical factor in larval rearing in hatcheries is the food supply and this occupies a large part of the effort. The food consists of very small diatoms and small flagellates about 10 μm or less in size. Culture of these algae to maintain a constant supply is not a simple task. Stock purity must be maintained. Also high growth rates are needed which requires strong light and a supplement of carbon dioxide. For more details on algae culture the reader is referred to Walne (1963) and Depauw and Pruder (1981) and the references in these publications.

Initially, the method adopted was to collect the seed on cultch in the hatchery. This produced problems, for growers were not prepared to accept newly set spat because of the small size. Furthermore, the handling and shipment of large quantities of cultch is expensive. To produce large seed the hatchery must maintain and feed the young oysters until the acceptable size of 5–10 mm in diameter is reached, and this is costly. An alternative is the production of single or cultchless spat by setting the larvae on very fine shell particles (300–400 μm) or on plastic sheeting from which the small spat may be readily removed. These spat, when 5 mm or less, are sent to the growers in large numbers in small packages (Fig. 37).

The grower has to grow the cultchless seed on trays to a size where they can withstand the vicissitudes of bottom culture — usually a size of 2 cm. Growth on trays creates other problems. Unless the trays are held in perfectly still water, spat tend to move to the corners in piles thus interfering with growth. If they are grown in still water, the spat of some species, especially *Crassostrea* species, become attached to each other and to the mesh of the trays unless they are stirred occasionally. Handling single spat involves many problems.

One solution to the problems of hatchery produced spat is the shipment of eyed larvae, almost ready to set, to the oyster grower who will set them on his own cultch in his own tanks. Two million larvae occupy a volume about 2 x 2 x 2 cm (matchbox size). These larvae may be stored (temperate zone) at 5° C for 5 to 8 days.

When the grower receives the eyed larvae, wrapped in a wet nylon cloth, they are placed in tanks where sea water, filtered through a 100 μm screen, has been heated, if necessary, to at least 25° C and maintained at that level for the duration of the set which may take up to 48 h. Clean cultch is already in the tanks. The larvae are added at the rate of 100 larvae/100 cm^2 of cultching surface (one side). Sets of up to 100 spat/shell are obtained.

Fig. 37. Hatchery bred single spat. a = shipping container. b = trayed single spat, 15 mm diameter. c = close–up of spat (from Quayle, 1988).

The larvae are distributed throughout the tank by bubbling compressed air from a system of pipes on the bottom of the tank. The spat are maintained in the tank until enough dissoconch shell has been formed to give a good attachment area (1 – 2 mm) which may take 3 or 4 days. When the cultch is removed from the tank to the open water will depend on its temperature. If the sea temperature is considerably lower than the tank temperature, the latter should be gradually lowered until the two temperatures are about the same. This will likely not be a problem in the tropics. If the spat are held in the tanks it may be necessary to feed them, either with food grown specifically for the purpose or with raw sea water. The seawater is coarsely filtered to remove the larger organisms which may be competitors or predators, leaving only potential food organisms.

At first glance hatcheries would appear to be the perfect solution to the seed supply but the capital investment in facilities and trained manpower make them almost a last resort. There are functioning hatcheries but probably more have gone out of business than there are operating. In North America and Europe there are only a few economically viable hatcheries. Accurate figures are not available but probably less than one tenth of one percent of the world's oyster production is from hatchery produced seed. Virtually all existing commercial hatcheries are producing temperate species. Control of the reproduction of tropical species is not as advanced as it is with temperate species. It might be said "hatcheries if necessary but not necessarily hatcheries."

CHAPTER 3

MEASURING GROWTH OF BIVALVES

Growth in molluscs consists of increases in both the shell and the soft body, but the easiest thing to measure is the shell. Since in most species there is a direct relationship between the linear dimensions of the shell and the mass, the changes in shell length are often used to monitor growth. As discussed in Chapter 1, the long axis of the oyster is biologically the height, since this is the anterior–posterior dimension (Fig. 4), but to an oyster grower it appears to be the "length" so the general practice is to use it as such. In practice the axis perpendicular to the length is called the width.

The weight of the animal would seem to be more important than the shell length, since it is the weight of meat (or shell produced, in special cases) that is of concern. However, weighing presents some difficulties. There are often drastic seasonal changes in the ratio of meat to shell weight as a result of the reproductive cycle. There is also sometimes fouling of the shell which adds to the oyster weight. When the oyster is grown on cultch it is difficult, if not impossible, to weigh during the growout period.

Growth in molluscs can be measured by changes in volume, since volume integrates length, width and thickness. This is particularly useful when there is considerable variation in the shape of the shell, as occurs in many oysters.

Shell growth seems to be primarily dependent on temperature, but energy is required, derived directly from food or glycogen reserves, to convert the calcium in sea water to shell. In temperate waters, there is little shell growth in most species when the water temperature is below about 10° C. This winter period is also the time of minimal food supplies, so molluscs are almost in a state of suspended animation. In tropical waters, shell growth is almost continuous with the consistent high water temperatures existing there.

Time of submergence is also a factor in growth. Intertidal oysters grow more slowly than those in suspended culture where they are continuously submerged. The type of ground on which oysters are grown also influences growth. Oysters grown singly on firm ground assume a round, deeply–cupped shape with much exterior shell fluting, while those grown on soft ground assume an elongated shape with smooth exteriors and long mantles posterior to the adductor muscle. The shape of oysters is so variable, depending on the immediate environment, that they might appear to be different species.

MEASURING GROWTH

Length measurements may be done either with an ordinary ruler, a measuring board, calipers, a marked graph paper, or by photograph. An ordinary ruler is probably the most useful instrument for measuring length or width in oysters. It is rapid and sufficiently accurate except for special purposes. A measuring board is simple to construct and to use (Fig. 38). Calipers are accurate, but unnecessarily so, are awkward to manipulate, and slow, particularly if the vernier is used. The graph paper method is useful in the field where recording is a problem. A sheet of squared paper is placed in a plastic envelope and on a board of equal size with one end blocked off by a 2 x 2 cm square rod. The oyster umbone is placed against the block, and the position of the bill or posterior end is made by puncturing the plastic and paper with a needle. These lengths may then be recorded in the comfort of the laboratory.

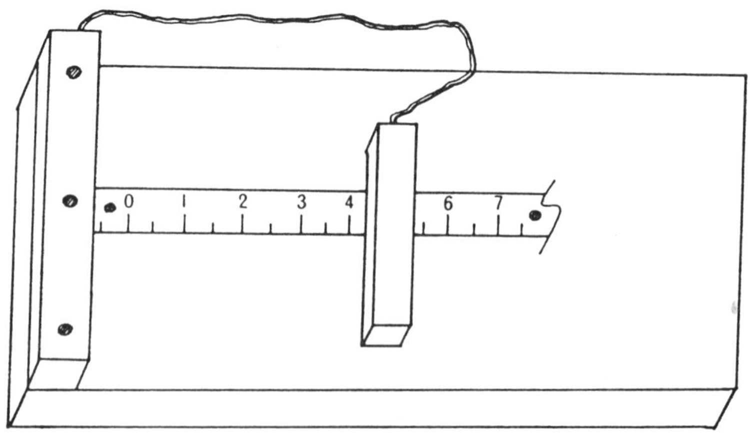

Fig. 38. Measuring board (from Quayle, 1980).

Photographs are useful (but expensive) when measuring oysters in sites as on a mangrove root. Care must be taken in making sure of the correct magnification when measuring the prints. A modification of the photographic method is photocopying if a machine is available. This is particularly useful for measuring small spat in the 2 to 10 mm range (but it is not handy for fieldwork!). The small oysters are spread out on a sheet of glass or plastic and exposed. Here again care must be taken to ensure the correct magnification.

Too often time is wasted with what is thought to be very precise shell measurements. Measuring very carefully to the nearest fraction of a millimeter on an oyster with an irregular shell margin is a waste of time. The measurements will usually be used to calculate a mean length or placed into class interval units in a length–frquency table or graph (Fig. 39). The variation in size of any sample of oysters is so great that reliable estimates of size are obtained by increasing the number sampled, not by increasing the precision of individual measurements. Thus, for most field work the measuring board or the ruler are the preferred methods.

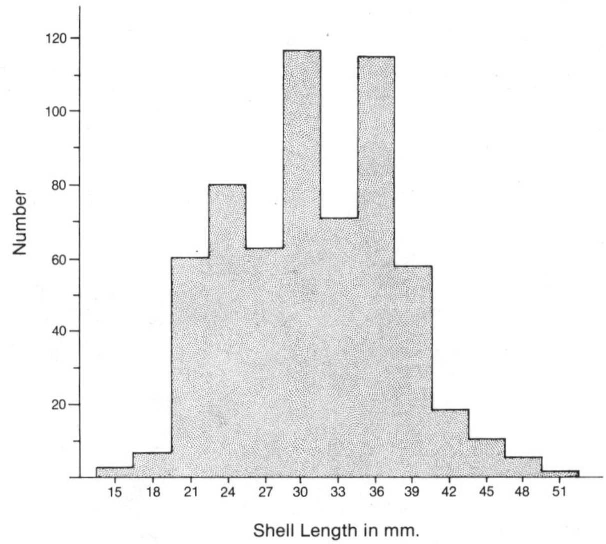

Fig. 39. Length frequency of Crassostrea rhizophorae *from Jamaica. The data were grouped in classes 3 mm wide. The mean length is 30.9 mm (from D. Thompson, unpublished).*

Another measure of growth is the increase in shell weight. To eliminate the meat weight variations, the oysters are weighed in water (see condition factor methods). This is a sensitive method, and if an oyster is not laying down shell in a regular way, it can be an indication the oyster is being disturbed in some way, such as with a disease. In temperate waters where linear shell growth is minimal or not at all, this method will indicate possible internal shell accretion. Experience will indicate how much of a problem the seasonal meat weight variation is. If the variation is not great, whole body weights, weighing in air, may be satisfactory. If air weights are used, care must be taken not to let the animals leak, that is, gape and allow the shell liquor (water) to escape. At the same time, it is desirable to dry the shell as much as possible of the water adhering to the outside. The principles about precision regarding length measurements apply to weighing. Measuring volume has already been described in the section on condition factor.

Growth Experiments

Growth in bivalves may be studied by any one of three methods or a fourth method particularly suited to clams:

1. Measurement of individuals from random samples of a population;
2. Successive measurements of marked individuals;
3. Measurement of annual growth rings, if their validity as indications of seasonal growth has been established;
4. Acetate peels of cut shells.

The first method is to compare successive length frequencies of a sufficiently large random sample (Fig. 40). The same group may be repeatedly measured, or a new group can be used providing it is from the same sampling site.

Random sample measurements are useful only when the breeding season is short so a new brood enters the population as a well defined group with a limited size range. If this is so, each year class appears as an isolated, or nearly isolated, mode in a size–frequency distribution. If the breeding season is an extended one, there will be an expanded range of sizes, and faster

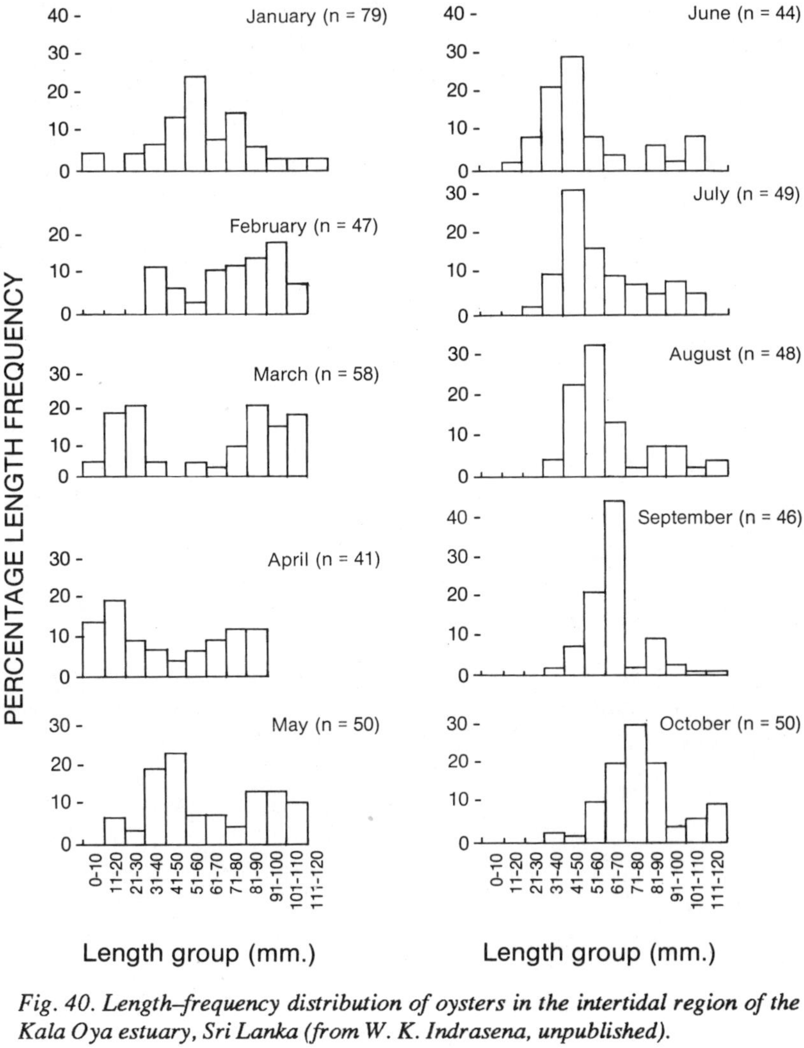

Fig. 40. Length–frequency distribution of oysters in the intertidal region of the Kala Oya estuary, Sri Lanka (from W. K. Indrasena, unpublished).

growing animals of one year class may be confused with slower growing animals of another. This limits the usefulness of this method except in special cases.

The second method is a positive direct method that makes repeated measurements of a single specimen over a time period. This requires the ability to recognize individuals and this is done by affixing a label, either by glueing a tag to one valve with waterproof glue or by drilling a small hole in the left valve umbo and tying a tag (Fig. 41). A number may be etched on the shell by means of an electric drill. Numbers written on the shell, regardless of the ink or paint used, usually wear off, but they might last if protected by a layer of transparent plastic such as some of the liquid glues. For some bivalves, like mussels and clams, a filed notch on the shell edge leaves a permanent reference point, in case the marking fails.

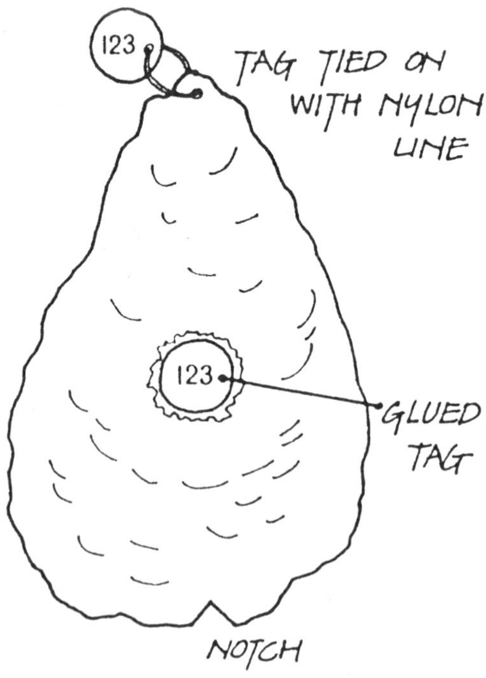

Fig. 41. Tagging methods (from Quayle, 1980).

A disadvantage of repeated measurements is the formation of a disturbance check each time animals are removed from their habitat and growth is interrupted. However, infrequent disturbances may not have a significant effect on total growth, so they may be ignored.

Measurements of annual growth rings (in seasonal environments where they are produced) provides an accurate method of estimating rates of growth. The animal is unmolested, other than by natural disturbances, during the growing period. While in most instances oysters are difficult if not impossible to age from shell markings, this is not so with most clams. This is based on growth interruptions which leave an abnormally deep line on the external surface of the valves. Growth interruptions may be caused by temperature change, usual in temperate zones, and these marks are called winter checks. In the tropics, cessations in growth are most likely caused by salinity changes. Removal of clams from their normal habitat may also cause growth interruptions called disturbance checks. One problem is to distinguish between seasonal and disturbance checks, but usually the latter are not as distinct as the former.

However, validity of rings on the shell as indications of age and seasonal or annual growth must be checked for each species. This is done by following the growth of marked clams. A preliminary gauge is whether or not measurements of growth rings on a few specimens show a typical bivalve growth curve shape when graphed (Fig. 42). Once the validity of annual rings is established, it is possible to compare growth at various sites. Seasonal growth may be studied by monthly random sampling of a natural population, or more accurately, with monthly measurement of marked animals.

For measurement of individual rings, some form of calipers with sharp points or edges is necessary. If calipers are not available, dividers are a good substitute. The divider distance may be readily transferred to a ruler or graph paper. The position of the ring most difficult to determine is the first ring, since this can be the most variable. If the new settlement is just before the cessation of seasonal growth, the animals enter the new growing season at a small size, sometimes not more than 2 or 3 mm. If they set well before the cessation of growth, they may enter the new growing season up to 10 mm in length.

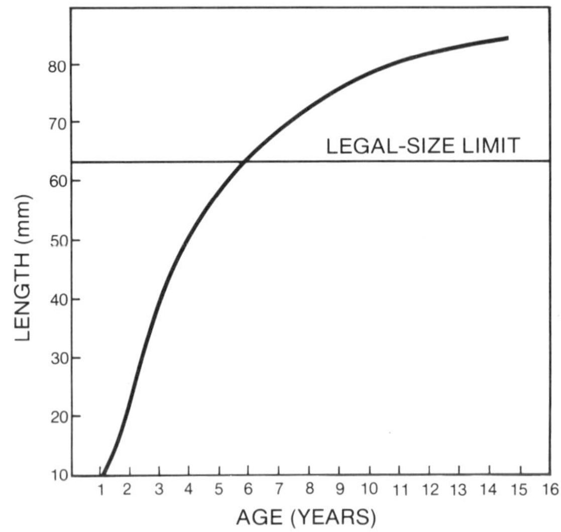

Fig. 42. Growth rate of butter clams from Seal Island, Strait of Georgia, B.C. (from Quayle and Bourne, 1972).

Growth rings also show internally on cross sections of the valve. For many species, a technique known as the acetate peel method may be used. Shell sections have been used for many years to age molluscs, but the acetate peel method is a relatively recent development. Instead of sectioning the whole valve, only the umbonal region is cut, for this contains a relatively complete record of growth on a lesser scale than the full valve section.

The method consists of making a cut through the umbone of a valve with a lapidarist's saw or similar instrument (not necessarily a diamond saw) that will make a clean cut, through the umbo or chondrophore, toward the ventral valve edge (Fig. 43). Small fragile shells may have to be embedded in epoxy to prevent shattering when being cut. The cut edge of the umbonal section is polished by flatwheels with various grades of diamond grit such as 180 and 600. Final polish is done with tin or aluminum oxide sand papers (0.3, 3.0 and 30 µm). Care is taken to ascertain the edges are square with no scratches on the surface.

Fig. 43. Enlargement of acetate peel from a clam, Mercenaria mercenatia, *showing the middle homogenous layer growth bands. Light regions in the photograph correspond to relatively opaque regions of the peel. Light bands indicate growth in years 1973 to 1977. Scale bar = 1 mm (from Fritz, 1982).*

The cut and polished shells are washed in a dilute soap solution and well rinsed. Scotch tape applied to the polished surface will remove any extraneous material that might have adhered. The next step is to etch the cut surface by arranging the valve in an upright position so the cut surface may be flooded with an aqueous solution of 1% hydrochloric acid for 1 to 2 minutes. It is then rinsed and oven dried at 65° C for 10 minutes.

When the shell is cooled, it is again set upright (usually in sand) and the polished surface flooded with acetone. Before the acetone evaporates, a film of acetylcellulose 0.076 mm. thick is laid on the cut surface, starting at one side to prevent bubble formation. The film is then flattened and smoothed. After about two hours the dry acetate peels are removed from the shell and mounted between two microscope slides that are taped together.

The slides are read with a micro projector (x50 magnification) but a microscope is satisfactory. Each peel should be counted more than once and care taken to differentiate true annular lines from cracks or scratches. The first ring is often difficult to determine.

The acetate film may be purchased from:

Transilwrap West Corporation
274 Harbour Way South
San Francisco, CA 94080 USA

Most important is the question of sample size. This should not be less than 100 and preferably close to 200 in the initial stages of the study. After the first few samples, the degree of variability may be assessed and the sample size adjusted according to statistical analysis.

With tagged samples, the environment for each specimen should be identical and the best way of accomplishing this is by means of a tray, whether suspended from a raft or on the bottom. If on the bottom, the tray should be held slightly off bottom to avoid siltation. An additional precaution in most growth studies is to notch the edge of the shell with a triangular file to provide a permanent reference point. The notch will remain on the shell for the life of the mollusc. This is particularly useful in molluscs with smooth, regular shells like mussels or clams.

In the tropics, where growth rates are rapid, studies may be completed within a short time. In areas of slow growth a composite growth curve may be established rapidly by studying the growth of several size groups at one time. For instance, one group of 1 cm in length, another of 5 cm and a third of 10 cm may be studied simultaneously. The three rates so developed may then be combined into a single curve. The smallest group will grow most

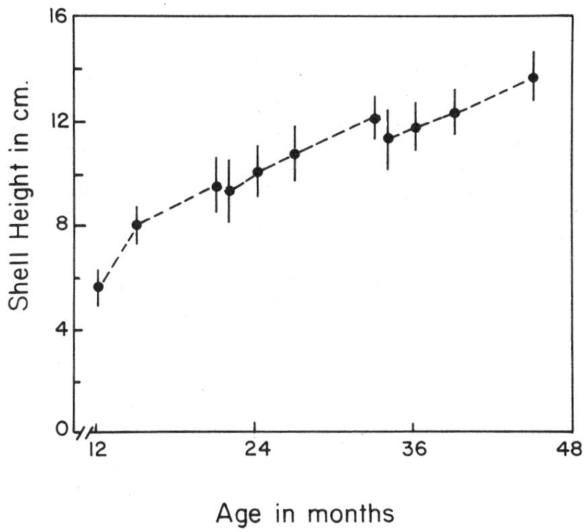

Fig. 44. Mean of shell height and standard deviation (vertical lines) for Pinctada margaritifera *in the Red Sea. Three year classes, indicated by the three data sets, were all measured in the same experiment (from A.G. Elnaiem, 1984).*

rapidly (Fig. 44) and the largest more slowly. This will approximate the growth only in the time period studied.

Growth in bivalves is highly variable, varying from season to season, from year to year, and from site to site even when close together. Because many groups, especially oysters, are difficult if not impossible to age in most instances, growth studies in wild populations are difficult. Thus, it is necessary to know the history of an animal, such as the time when it was spatted and the environment during its growth period.

Growth Relationships

Maximum length is the most used measurement of growth in bivalves, but height, thickness and weight should also be included in any growth study. Along with total weight, meat weight in relation to length should be measured to determine the size and age of greatest meat productivity. This, in turn, requires information on seasonal changes in condition factor. This may be determined for most bivalves in the basic way used for oysters, although care must be taken if the species gapes allowing air to enter.

A useful source of information for the grower is the relationship between the various measurements of growth. Of particular interest is that between size of the animal and its meat content. About 300 single well–shaped oysters produced the data shown in Table 2 and Fig. 45 at the beginning of November. The condition factor was not particularly high, with little recovery after spawning. A few had not spawned and were still in a ripe condition. It will be noted there is little change in the meat weight after the oyster has reached a length of about 100 mm (4 in.). The absolute weights

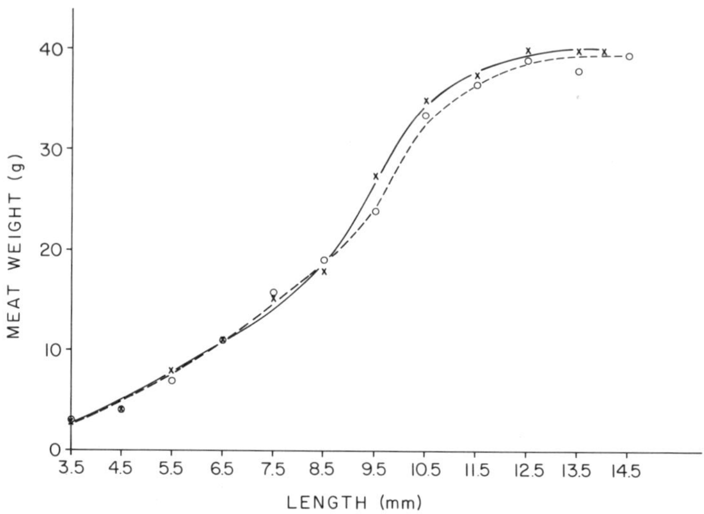

Fig. 45. Length - wet meat weight relationship of Pacific oysters. Hammond Bay. November, 1985 (from Quayle, 1988).

as shown here have little value in themselves since they change throughout the year, but the relationship remains valid for there is a direct relationship between meat weight and internal shell volume as shown in Fig. 45. Thus, many small oysters can yield as much meat as fewer large oysters. For instance, 100 oysters 100 mm (4 in.) in length will yield as much meat as 90 oysters 145 mm (6 in.) in length (Table 2). The economics of this situation is the balance between the extra cost of shucking more small oysters against the cost of holding the oysters for the time required to grow an extra 45 mm (1.8 in.). As mentioned elsewhere, theoretically, small oysters could fetch a better price than large oysters and it is almost certain the market acceptability would be greater. It would be of value to growers to examine their own oysters in this context.

Similarly, when studying the effects of various planting densities, the questions to be answered will decide which parameters are to be measured. Recording only shell growth neglects biomass (meat content) whose production is the object of the culture exercise. It may be that in a given space many

Length (mm)	Whole weight (g)	Internal volume (cm^3)	Wet meat weight (g)
30–39	40	2	3
40–49	38	5	4
50–59	45	7	7
60–69	53	13	9
70–79	101	24	16
80–89	119	28	19
90–99	142	34	24
100–109	181	46	34
110–119	222	56	36
120–129	246	60	39
130–139	240	53	38
140–149	275	55	39

Table 2. Size relationships in the Pacific oyster *Crassostrea gigas* from Hammond Bay, British Columbia, Canada (source: Quayle, 1988).

small animals will provide a greater biomass than few large ones. Since large animals result either from a longer growth period or from a lower stocking density (a larger culture space), the production of large animals may not be the optimal or most economical use of space or time, whether using trays, strings or bottom culture. The seasonal aspect must also be considered, for meat content varies according to the reproductive state (ripe or spawned out) and success or failure of the fattening cycle. Earlier harvesting of animals in good condition may be better than later harvesting of larger animals in poor condition.

The market requirements and processing methods may also determine whether the total biomass is more important than individual size. For example the clam *Tapes philippinarum* in Japan is canned (meat only) at a shell length of 20 mm or less, while in North America (west coast) the minimum size for the same species for utilization as steamed clams (sold in the shell) is 40 mm or more.

In summary, growth data should provide the following relationships:

1. Length-age
2. Length-height
3. Length-thickness
4. Length-total weight
5. Length-shell weight
6. Length-meat weight
7. Shell weight-meat weight

A simple method for transforming the normal length-age growth curve to a straight line is to plot on graph paper the length at one age against the length at the next age of growth (e.g., length at year 1 against length at year 2, and year 2 against year 3 and so on) Walford, (1946). Normal growth for the butter clam (*Saxidomous giganteus*) and the straight line transformation are shown in Fig. 46 as well as that for the same species at a point along the coast 300 miles to the north. These can be described by the notation l_1, l_2, l_3 and by comparing

$$K = \frac{l_3 - l_2}{l_2 - l_1} \quad \text{or} \quad \frac{l_4 - l_3}{l_3 - l_2}$$

a constant K can be calculated. When the constant is positive and less than one, the yearly growth increments decrease. Also, the ultimate length of the species can be calculated from the formula:

$$l\ (\text{ultimate}) = \frac{l_1}{1 - K}$$

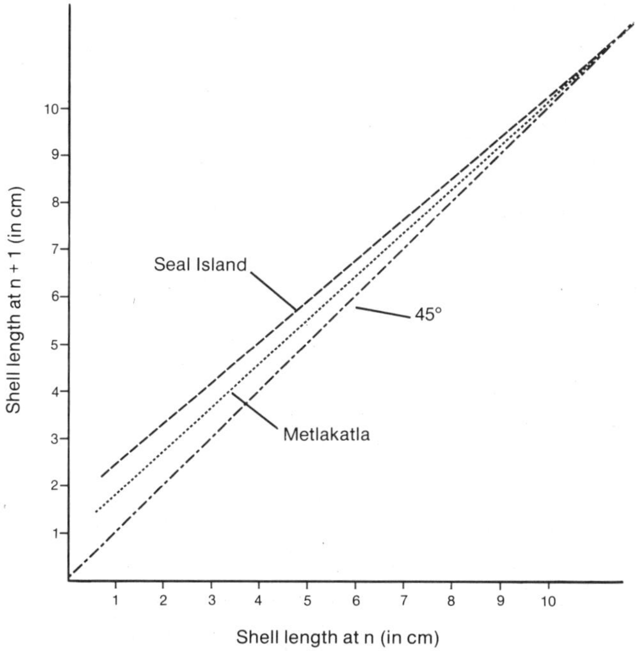

Fig. 46. A Walford Plot of shell lengths of the butter clam, Saxidomus giganteus, *from two different locations. The 45° line is shown with dots and dashes. The two regression lines are above the 45° line and intersect at the theoretically maximum shell length.*

It may also be determined graphically as the point where the length at age n equals the length at age n + 1, which is where the transformed line intersects a line drawn at 45 ° from the zero point (Fig. 45). Thus, it is possible to make comparisons between individual or population growth patterns. This is particularly useful for molluscs whose age can be determined from growth rings.

MODEL EXPERIMENTAL PLAN

With some hesitation, a "model" experimental plan is presented to show the kind of studies that could be done in a situation where one is exploring the culture potential of one or more indigenous species. For our purposes here, it is assumed that there are one or more species of oysters that show potential either because they are already being exploited as a fishery, or the existing populations contain animals of reasonable size compared to marketable animals elsewhere. This plan does not include an assessment of potential markets. This is something that must be done in a preliminary way before fieldwork is started and expanded if the field work shows potential for production. Oysters are used as an example and only the details of culture methods need be changed for other species.

The hesitation in presenting a model is out of concern that the "model" will become a blueprint. The procedure given will not be completely or in part applicable in all situations. Careful thought must be given to the details of every new situation. It is hoped that the model will give an idea of the scope of a development project and a feeling for the timing of various activities.

The first experiments are done with single oysters held in trays to determine the growth curve of the species at the selected site(s) under good, if not ideal, conditions. It is started with oysters of different sizes, so a picture of the whole growth curve can be obtained quickly after the pieces are put together. The second experiment is done with oysters on experimental cultch to determine the growth curve on cultch from setting.

Individual oysters should be separated from the mangrove roots or other natural substrate. If there is more than one species present they should be identified and kept separate. Those that are not "obviously" one or the other species can be kept as a third group. The differences will be clearer in the larger oysters. It may not be possible to separate the species in the smallest size class.

It is suggested that a start be made with animals in the following size classes: 5, 25 and 50 mm. If the 5–mm group is difficult to separate or to hold on trays a 10–mm group can be used. The oysters should be removed from the substrate and held in the trays for one week to check for mortalities. Then the measurements can be taken. The size classes should average about 5, 25 and 50 mm, but there will of necessity be some variation in size within the groups. The range of sizes in each case should be kept as low as possible. For example, no oysters should be outside the following ranges: 3–7 mm for a 5 mm average, 7–13 mm for a 10 mm average, 22–28 mm for a 25 mm average, and 46–54 for a 54 mm average. These are guidelines. It is more important that the required minimum number of oysters be maintained (150/size class) than to stay strictly within these ranges.

The oysters should be measured monthly, but the trays should be cleaned weekly. As the oysters grow, they may need to be thinned and put in different trays. If they grow enough, they should be put on the next larger size mesh available. The mesh of the trays will severely restrict water flow as the mesh becomes fouled. It is important to be sure that the trays stay clean, otherwise the results will reflect a slower growth rate than is really possible. It is probably wise to use a larger mesh as a cover than is used on the bottom. The bottom mesh must hold oysters of all sizes. The top mesh is a safety net to avoid spills and to prevent the entry of predators. The top does not have to hold all oyster sizes.

The trays should be held on a raft, rack or longline at or near the place where oyster culture is proposed. Once the growth curves of the three size classes overlap (the small size class has grown to an average over 25 mm and the 25 mm class is over 50 mm), the experiment can be terminated. These oysters can then be used for measuring condition factor. They will probably be of a regular shape and thus easy to handle. This experiment should be repeated after the first trial, perhaps after 6 months.

At the same time growth studies should be done on oysters that have been collected on cultch and grown out on the cultch as a comparison to the growth studies in trays. A convenient cultch for these experiments is pieces of asbestos 10 cm x 10 cm cut from sheets about 10 mm thick. (This is not recommended as cultch for commercial culture but is excellent for experimental work as the size can be standardized and they can be labeled easily.) The growth experiments using the asbestos plates can start at any time the spat is available. The spat could come from any station but should be held at stations to be used for culture.

The objective of these studies is to determine the growth rate of the oysters and to determine what differences exist between species and sites. If there is any indication that different species are being collected at different sites, the collectors with these different species should be placed at the various sites for growth studies.

The purpose is to follow 5–10 oysters/plate through to market size. To allow for mortality, there should be more than 10–20 oysters/plate at the start. After the plates are examined for spatfall, they should be returned to the water for two weeks. If the spatfall is heavy, they should first be thinned so there are 20–50/side. After two weeks the plates should be cleaned and the oysters thinned again, so there are 10–20 widely spaced oysters on a side. One corner of the plate should be cut for orientation. Each plate should be numbered and each oyster circled with a pencil to distinguish it from spat that might set later. A diagram of each side with oysters should be made on the top of the data page (see example, Fig. 47). Each oyster should be assigned a number so its growth can be followed. The initial shell height and length should be recorded. The plates can then be placed at the appropriate site. They should be cleaned every fortnight, and any new spat or fouling organisms removed. The oysters should be measured every 4 weeks (monthly).

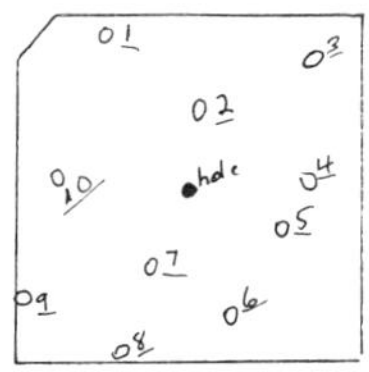

Sample Data Sheet

Station ________________

Plate ________________

Started ________________

Oyster #/dates

	l	w	l	w	l	w	l	w	l	w
1										
2										
3										
4										
5										
6										
7										
8										
9										
10										
mean										

Fig. 47. Sample data sheet to be used to monitor growth of oysters set on experimental plates. The plate for this data sheet is diagramed in the upper right corner for reference. One corner of the plate is cut for orientation. Two sides can be monitored with the same or different data sheets.

If the oysters become crowded, some should be removed to avoid over growth. They should be removed randomly with respect to size. One point that needs to be reiterated is the randomness of eliminating oysters as they become crowded. The most objective way to do this is to consider two adjacent oysters at a time. Temporarily, label one oyster #1 and the other #2. Flip a coin to determine which one should be eliminated. If you systematically remove the small oysters the resulting growth curve will be biased.

These studies should continue for as long as possible. Each month the mean size can be determined and plotted to show the growth curve. If this is done immediately, any site differences will be revealed and the handling of future sets of growth plates may be adjusted either as to site, level, measurement frequency, etc.

It is important to identify the species in these growth trials if more than one occurs in the area. As the oysters grow and have to be thinned, those removed can be examined to determine species. These should not be the smallest oysters on the plates. They should be randomly selected. Selectively removing oysters (large or small) will produce artificial "bumps" in the growth curve and give a biased idea of the growth potential.

STATISTICS

When starting a project to evaluate the potential or to actually establish bivalve culture it is usually necessary to carry out some types of trials. The trials will be to test new sites, compare culture methods, evaluate alternative species, etc. Most of the results will not be so clear cut that the answers to questions posed are obvious. More than likely the results will be in the form of showing that oysters grown at a certain depth are bigger than at another depth, the mortality is lower at one site, etc. In any case, the outcome will be measurements of some nature (sizes, yields, etc.) which will be compared to arrive at some conclusion. The problem is to determine whether the observed differences are big enough or consistent enough to be worthwhile. With such data it is possible to carry out statistical tests in order to make the conclusions more sound.

In this text we will not try to give an introductory statistics course. Rather a few basic points will be emphasized. To fully understand these points one must have some understanding of statistics. A researcher must have an understanding of the statistics that will be used in the project. Most statistical tests require some assumptions about the data and too often the assumptions are ignored in practice because the user does not stop to remember what was learned in the statistics course. These assumptions should be understood by the person doing the research. It is advisable to consult a statistician if the analysis to be used is not fully understood. Such consultation should be obtained before the experiments are started as it may be necessary to change the experimental design to obtain the best results.

The statistics used will vary with the kind of data and the purpose of the work. The first thing one normally does is to calculate descriptive statistics. The mean is the most common statistic used. The mode and the median are also useful sometimes. An understanding of the variation in the data is essential in statistical analysis. Without knowing the variation in the variables few statistical tests can be done. The most common statistic to measure variation is the standard deviation (the square root of the variance). The range is informative but of limited value in statistical tests. Formulae for these statistics are given in Table 3.

$$X = \frac{\Sigma x_i}{n}$$

$$s^2 = \frac{\Sigma (x_i - X)^2}{n - 1} = \frac{\Sigma x_i^2 - \frac{(\Sigma x_i)^2}{n}}{n - 1}$$

$$s = \sqrt{s^2}$$

Table 3. Statistics. n = sample size: X = mean; x_i = individual x; s^2 = variance; s = standard deviation.

Before any tests are done on the data, the descriptive statistics should be calculated and studied. The mean and standard deviation of each treatment or group should be tabulated or plotted. Often plotting the data is useful in getting an appreciation of what is going on.

A simple test for the difference between two means is the t-test. A statistical text should be consulted for the proper use of the t-test. Such things as checking the equality of the variances in the two groups should be done. There are different formulae for different situations. Do not just use the first or the simplest one!

When there are more than two means, the appropriate analysis is probably an analysis of variance rather than multiple t-tests. If the question is one that implies trends or associations then regression or correlation analysis would be applicable. In all cases the researcher should have a good idea of what kind of statistical analysis will be done at the end before doing the experiment.

The growth of bivalves is very variable. Animals grown on the same tray will show large differences in size. This may be due to different microenvironments even within the same tray (for example, accessability to waterflow) or competition among individuals. Variation from tray to tray, with depth, from site to site and from year to year will also be large. With all this natural variation, the job of demonstrating statistically significant differences due to a treatment is made more difficult. Thus, it is necessary to have an adequate experimental design.

An elementary consideration in planning the experiment is to be sure the sample size will be large enough. Most of the statistical tests used employ the standard deviation or variance in determining whether there are significant differences among means. The structure of the formulae for these tests show that, for a given difference in the means, it is easier to detect a significant difference if the variance (or standard deviation) is minimized. Of course, we cannot change the variance in the population sampled, but we can minimize our estimate by increasing our sample size, n. This result is obtained because the sum of squares is divided by n - 1. Another way of looking at this is to think of it as increasing the accuracy of our estimate of the variance by taking a larger sample.

Speaking of accuracy, it is worthwhile to consider the distinction between accuracy and precision. Accuracy is a measure of how close the estimate is to the true value. Think of it as how close you are to hitting the center of the target. Precision on the other hand is how consistent you are in repeated measurements (or shooting at the target). If the vernier calipers used to measure oysters are in error (short 5 mm) the measurements may be very precise but consistently inaccurate. A common concern for precision is shown by the insistence on using vernier calipers to measure oysters. The calipers supposedly give a more "precise" measure of the length of an oyster than a ruler which can only be read to the nearest millimetre. However, especially with oysters, when the shell margin is often very irregular, a length measurement to a fraction of a millimetre is "too precise." If the measurement is taken a few millimetres to either side the value will be quite different.

Furthermore, taking, recording and analysing many decimal places in a measurement (length, weight or whatever) takes time. That time is probably better spent measuring more individuals, that is, increasing the sample size, n. In this way a better, more accurate, estimate of the mean will be obtained. The errors due to measurement (where the ruler is placed, whether there is a little water on the animal when it is weighed) are usually small when compared to other sources of variation. This is not to say that the measurements can be sloppy! It is just that there has to be a balance between the precision of the measurements and the time involved. Sometimes precision is only apparent while accuracy is not being improved.

The biological value of the measurements should be considered. Rarely will it be worthwhile measuring a juvenile or adult bivalve any more precisely than to the nearest millimetre or 0.1 g. Although the calipers or the balance can measure to more decimal places, the "precision" gained is usually of little significance. One way of checking this is to do repeat measurements of identified individuals. This is best done by a second person who does not have the previous data in front of him or her. This should be done with a sample of 30 or more animals.

In experimental work of the nature discussed here, comparisons are usually made. In order to do this we have to have more than one group. If we are testing a new culture method, we call the standard technique the "control" group. However, not all experiments have a control *per se*. But in all cases where methodology is changed, a control must be present with which to compare the results of the new method. The control should be run at the same time as the experimental groups, with the only difference being the variable being evaluated (depth, fouling, density, etc.). Otherwise, sources of variation other than the factors being evaluated may cause differences between the groups. Too many times people working with bivalves change several things at one time and are pleased that the new approach works better. Unfortunately, if two variables have changed, one may have improved the situation and one made it worse but the combined effect is better. Or a third variable may have improved something like primary productivity while the two manipulated variables actually made things worse! Without step by step evaluations the maximum improvements are unlikely to be made.

In planning an experiment or an evaluation, it is imperative to have a clear plan. Resources are always limiting whether time, material or money. The experimental design must use these resources efficiently, yet still be able to provide the answers needed. The first step is to state clearly the objective of the study. This will set the framework. Then an hypothesis is presented that can be tested. Statistical analysis of the data then determines whether the stated hypothesis will be accepted or rejected. If an hypothesis is rejected, it does not necessarily mean it was "wrong." It simply means we do not have enough evidence to accept it. If we think the hypothesis is worth evaluating again, the experimental design should be reviewed. Again, we will not go into any detail here. Texts should be consulted for a complete discussion of hypothesis testing.

Replication of experimental units is very important. As we said earlier, there are many sources of variation. Individual bivalves have different innate growth rates because they differ genetically and physiologically. Microenvironmental differences exist even from animal to animal. Trays and strings may differ because of their unique location or treatment. Thus, if different growth rates are observed in two trays that have different densities, how does one separate the effect of stocking density from the microenvironmental differences between trays? If the experiment is replicated so there is more than one tray at each stocking density, and the average of the trays at the two densities still differ, we can begin to suspect that the differences are due to the stocking density. The number of trays in a case like this is as important as the number of oysters in the tray. Two trays with hundreds of oysters in each still give only two means. If the variation from tray to tray is great, we will need more than two trays at each density. Usually there is information available that can be used to assess the experimental design. If not, preliminary experiments may be needed.

CHAPTER 4

UNWANTED SPECIES

As in terrestrial farming, the bivalve farmer must deal with a number of unwanted species. These vary from the nuisance organisms that grow around and on the desired species, to disease causing organisms, to those that prey on the farmed species. The first group of plants and animals are called fouling organisms. (On land they would be called weeds.) Usually they are simply a nuisance but at times they cause mortality. There are some diseases of which a bivalve farmer should be aware. Little can be done to cure the animals but culture procedures may be changed to avoid or minimize the effect of disease. There are also predators to be considered.

In this chapter, fouling, parasites, disease and predators will be discussed. The major groups that affect bivalve culture worldwide are discussed along with methods of control.

FOULING

Fouling, or more properly, biofouling is the attachment of marine organisms, either plant or animal, to the object of interest, whether it be oyster cultch or a boat. In oyster culture, it may be more of a nuisance than a serious problem, so its importance may be overemphasized. There is a tendency for inexperienced oyster culturists to become unnecessarily alarmed over a level of fouling that is, in fact, inconsequential. Also, it is often more economical to accept whatever deleterious effects of fouling there may be than to undertake costly control measures.

However, fouling can be a major problem in some areas. A rough rule of thumb is that there is difficulty when the volume of fouling approaches the volume of oysters or cultch. Depending on the market for which the oysters are being grown, the type of fouling organisms may also be a factor. Fouling is not as important for shucked oysters as for those used in the half–shell market.

Fouling is a problem with oysters because it may cause mortality, particularly in seed, reduce growth rate and, in suspended culture, cause flotation problems. It may seriously affect seed collection by reducing the surface area of the cultch, and some fouling organisms can feed on oyster larvae. The deleterious effects of fouling of cultch is one of the reasons for spatfall prediction.

Fouling is generally less of a problem in intertidal culture than in subtidal or suspended culture, especially in the tropics where the high air temperature inhibits the survival of soft–bodied invertebrates such as tunicates or anemonies, two of the main suspended culture fouling organisms. These organisms, among others, generally do not occur in low salinity waters as they are stenohaline, being restricted to a narrow, high salinity range. On the other hand, oysters, particularly the genus *Crassostrea*, are euryhaline, able to live and grow in a wide range of salinities. The genus often breeds best in areas with medium salinity such as 20 – 25 ppt. To take advantage of the euryhalinity of oysters and the stenohalinity of most fouling organisms, seed can be collected and grown briefly in high salinity waters and then cultured upriver in waters of lower salinity (15 ppt) if these are available nearby.

The main fouling organisms causing problems are sponges, anemonies, hydroids, bryozoa, tube–dwelling polychaetes, barnacles, mussels, tunicates and algae. Also in tropical waters another byssal attaching mollusc is *Isognomon*, and there are cementing species such as *Chama and Spondylus*.

Sponges

Sponges may be encrusting or solitary and erect, the former are more difficult to detach than the latter. In addition, one group (*Cliona*), bores into calcareous shells creating a honeycomb of tunnels. Some may penetrate the nacreous layer causing the oyster to expend energy in producing more shell to cover the penetrations. Tunnels may so weaken the shell that shucking becomes difficult. Some encrusting sponges may be controlled by exposure to sun in the early stages of development.

Anemonies

Anemonies are coelenterates, relatives of the jellyfish and corals, with a soft body and a tentacled oral surface. They are all soft bodied with a broad holdfast difficult to dislodge. They vary from about 1 to 15 cm in diameter. The soft body makes it unable to withstand exposure to sun for any length of time.

Hydroids

Hydroids are also coelenterates but with a peculiar life history. The sexual phase is a small jellyfish (medusa). This produces a planktonic larva which, after settlement, develops into a branching bush-like hydroid which, in turn produces the jellyfish asexually. The hydroids assume many shapes and sizes, but the typical fouling species (*Obelia*) may attain a length of 30 cm. The hydroids may be killed by sun exposure or may be removed manually.

Bryozoa

Bryozoa are also termed polyzoa or moss animals and may be encrusting or erect with branches. In both cases they are colonial, each individual being of almost microscopic size. The encrusting types are generally less than 1 mm in thickness, but one colony may completely cover an adult oyster shell. They are not harmful to adult oysters but may overgrow spat less than about 1 cm in length. The erect branching forms are seldom more than 3 or 4 cm high and are generally not of great concern.

Tube Worms

Tube worms are sedentary polychaete annelids whose tubes may be calcareous, parchment-like, mud or sand. They are not usually a major problem, but at some sites and in some years there may be excessive outbreaks that may require drastic action. Air exposure of 24 – 48 h may be effective for most species. Another tube worm of importance is *Polydora*, the mud blister worm (Fig. 48). *Polydora* burrows right into the shell edge from where they are readily visible. The tube is partially filled with mud and

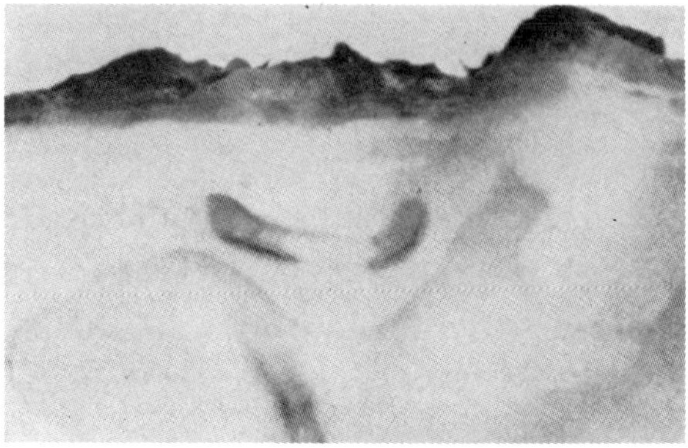

*Fig. 48. Burrows of the boring worm (*Polydora*) x 2 (from Quayle, 1988).*

sometimes the tube breaks through the nacre and the oyster is forced to lay down another layer of shell to cover it. *Polydora* is most often a problem in bottom culture in muddy areas. An alternative is off–bottom culture, but a brine dip may be useful for both it and the boring sponge.

Barnacles

One of the most important fouling organisms is the barnacle (a crustacean). Often there is more than one species, but the species determination is difficult from external characteristics alone, so some knowledge of barnacle anatomy is required.

There are two main types of barnacles, one with a fleshy stalk called goose barnacles and a sessile one with direct adherence to the substrate called acorn barnacles; the latter is the more common (Fig. 49). The acorn barnacles, the ones that the oyster culturists will encounter, have the body encased in a calcareous sheath composed of six overlapping plates cemented together (Fig. 49). The opening to the exterior is covered by four hinged plates, the anterior pair called the tergum and the posterior pair the scutum. It is through these plates that the feeding limbs extend. The main barnacle identification features are the way the plates are joined and the shapes of the tergum and scutum, although the mouth parts may also be important. For the non-specialist, the shapes of the tergum and scutum are characteristic enough

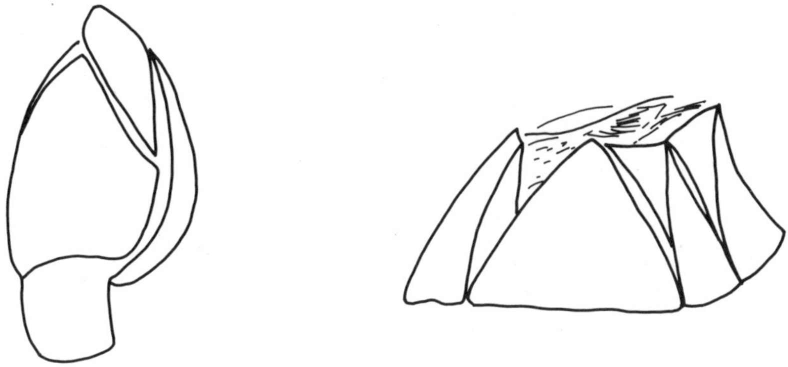

Fig. 49. Two types of barnacles: a) goose barnacle and b) acorn barnacle.

to separate most species. The barnacle is simply placed in bleach (Chlorox, Perfex) for 1 h and the tergum and scutum can be separated readily and examined with the stereoscopic microscope. The species may be temporarily labelled Barnacle A and Barnacle B.

The acorn barnacle, chiefly the genus *Balanus*, is probably the most ubiquitous of fouling organisms, occurring from the high intertidal to considerable depths and in great abundance, with settlement rates up to 25/ cm^2. Fortunately, most of them do not grow more than 1 cm in diameter. Often they breed at the same time as the oysters and thus can seriously foul the collectors. Some species attach with a solid plate, others only at the margins. There is also one group of barnacles that live within the shell of molluscs. Usually more than one species occur in a given area. Therefore it may be necessary to determinethe breeding season and distribution of each species so these may be taken into account in any scheme of mollusc culture and seed collection.

Mussels

Mussels are second only to barnacles in importance as fouling organisms, not so much in the tropics (except in certain areas such as Singapore) as in the temperate waters where the setting intensity can be very high. The growth rate is rapid and, combined with the numbers, this can cause a high level of competition with oysters for food, as both are filter feeders. Oysters may also be smothered. Knowledge of the breeding season and vertical distribution may assist the culturist in dealing with the problem by correct timing of culture activities. Probably the most economical method is exposure to sunlight when the young mussels are less than 5 mm in length. This becomes a problem if the mussel breeding season is long. Alternative sites may be a possibility if an area with reduced mussel breeding potential is found.

Tunicates

Tunicates or ascidians are either solitary or colonial. The solitary types assume many shapes but all adhere to the substrate by a broad holdfast. The main body is contained within a test or envelope of varying consistency according to the species, but most may be considered soft bodied and therefore susceptible to sun exposure. They may vary in size from 5 mm to 10 – 12 cm in length. The colonial type of ascidians consists of numerous small tunicate bodies encased in a fleshy encrustation up to 1 cm thick and capable of covering a whole adult oyster and more. Some species (*Halocynthia*) are edible and are both fished and cultured.

Algae and Bacteria

Bacteria and diatoms can form a thin layer of slime, of no concern except on the cultch, because on growing oysters they may be a source of food. If the slime film on cultch is heavy enough it may prevent spatting.

Larger algae such as *Ulva*, *Entermorpha* and, in some areas, the larger brown algae such as *Laminaria* may occur. These usually grow spread out in the water so their effect looks worse when the oysters are taken out of water and the algae covers the oysters. The large algae can add significantly to the weight in suspended culture and, thus, add to the flotation requirements. The filimentous and low, mat–forming types can reduce water flow through trays but are not much of a problem in string culture. Air drying is the usual remedy.

Species such as *Ulva* and *Enteromorpha* may cause problems in bottom culture if they develop thick mats on the oysters. As long as the algae lives there is usually no problem. However, if the weed dies, the rotting material may stress or kill the oysters. If there is a good tidal exchange, there will be little damage.

FOULING CONTROL

The three main categories of fouling control are physical, chemical, and biological. The physical methods include direct heat (flame or sun). Chemical methods use fresh water, hot water or chemical solution dips such as brine or copper sulphate. Biological control involves knowledge of the biology and ecology of the fouling organisms so the culture operation may be timed or placed to circumvent or minimize the fouling.

Physical Methods

These methods require removal of suspended culture oysters, strings, trays or sticks from the water. This is a costly operation and undertaken only as a last resort when the fouling becomes excessive. Probably the most drastic of these methods is the application of heat by some type of flame thrower. A delicate balance is required to ensure death of the fouling organisms and not the oysters.

Another method is the application of water with high pressure pumps and special nozzles. Aside from the handling costs, one problem with this method is that the fouling organisms may not be killed and are washed back into the sea where they can reproduce.

Probably the most frequently used method is air drying in the sun. It is most useful with soft bodied animals such as tunicates and anemonies, particularly young animals. This method is successful only with very young barnacles. Usually one full day's exposure is sufficient, but the time depends partly on the air temperature, humidity, wind, the size and type of foulers, and the size of the oysters. There should be initial experimental trials with small quantities to determine the optimal timing. The fouling organisms may not be removed immediately but may drop off several days afterwards. With small culture operations, many of the organisms such as mussels and hydroids can be removed by hand.

Chemical Methods

Chemical methods also involve removal of the oysters from the water, with the additional task of placing them in tanks containing the chemicals. Fresh water may be classed as a chemical to marine organisms and only 1 or 2 h of exposure is necessary to kill most fouling organisms, particularly the soft–bodied species. Hot water, owing to the cost, is almost out of the question except under special circumstances.

There are a number of chemicals toxic to marine organisms, but many are expensive, often have limited effect, and have so many dangers that they must be used with caution. Probably the most useful is copper sulphate in a concentration of only 1 or 2%. Care must be taken in its use relative to other organisms such as fish. A relatively short exposure time of 1 hour or so is adequate for most organisms, but here again preliminary experiments with small quantities should be made. Among the least expensive chemicals is brine; the bath should contain as much salt (NaCl) as will dissolve. In general, chemical control is impractical except in special circumstances.

Biological Methods

The biological method requires a knowledge of the life history and ecology of the fouling organisms in the culture area with special emphasis on breeding season, vertical and horizontal distribution of larval settlement, growth and survival. This entails a specific fouling study for the experimental collection of fouling organisms. A properly designed experiment may demonstrate weak links or sensitive stages in the progression of events that can be utilized to advantage in control.

Information such as the breeding season of a barnacle species may indicate that the cultch can be put out after the barnacle set, providing the barnacle setting occurs earlier than that of the oysters. In some circumstances, as in suspended culture, strings may be exposed purposely to collect the barnacle set, which in turn prevents settlement of more deleterious fouling organisms. In this case, the oyster seed should be large enough to prevent overgrowth by the barnacles. Knowing the vertical distribution of a fouling organism, it may be possible to place oysters for growth or the spat collectors above or below the fouling zone whether in suspended or intertidal bottom culture. This will require knowledge of both settlement and survival of the fouling organism. For example, as previously mentioned, in estuarine areas where there is a significant salinity gradient, it may be possible to collect seed in high salinity areas where breeding takes place but where heavy fouling occurs. (Indeed, in some contexts an oyster may be considered a fouling organism itself.) Then seed may be moved up the estuary to a lower salinity area where fouling is absent or minimal but where oysters still grow. This ability to thrive over a wide range of salinities is a characteristic of the genus *Crassostrea*, fortunately for the oyster grower.

Another possibility in the tropics, where oyster growth is usually rapid, is to culture in the off season for fouling if such a season exists. It may then be necessary to harvest at a smaller size as a compromise to evade fouling problems.

FOULING STUDY

Where bivalve culture is being developed and background information is lacking, fouling should be part of the culture study process by keeping records of time of occurrence and extent of fouling as it occurs on seed collectors or growing bivalves. In addition, specific studies may be carried out by exposing test materials of various types for different periods of time at various depths and at several sites. Preferably the test material should be oyster shell, or whatever material is used as cultch, or both, since these will be the substrates encountered by the fouling organisms. In some instances, particularly if quantitative measurements are to be made, it may be advantageous to use an inert substance that will provide a specific and consistent surface area. Such a material is asbestos cement sheeting about 5 mm thick and cut into 10 x 10 cm squares, an area which can be covered by the field of a stereoscopic microscope at low magnification. Wooden panels also provide information on marine borer attack but suffer from disintegration if the attack is severe or if the panel is left exposed for a long period.

Exposure of test plates can be intertidal, at a fixed tidal level such as from a dock, or variable as from a raft where depth intervals remain the same but change with reference to the mean low tidal level. The appropriate experimental design will depend on the type of culture.

Exposure designs are numerous and may be quite complicated as there is interaction between the several variables such as time, depth, site, season, materials, etc. A fairly simple method, especially for use in the tropics where there are essentially two seasons, a wet and a dry one, is to use a set of six numbered panels. These may be attached to a simple frame such as a piece of wood or metal from which they may be easily detached. Panel number 1 is removed and replaced at monthly intervals or whatever time period is desired. Panel number 2 is removed at the end of the second month, number 3 at the end of the third month, and so on down the line. Thus, monthly and increasingly lengthy exposures up to 6 months are obtained. Initial exposures and number of test panels are timed to cover whatever period of the year is deemed critical, such as rainy or dry seasons. The time interval might be shorter where fouling is heavy or growth is rapid.

Fouling organisms may be studied quantitatively by measuring weight, volume, percent cover or in some cases by numbers/unit area as with barnacles. Without adequate literature, identification may be difficult but the important groupings such as sponges, tunicates, etc. can be separated. Species may be temporarily identified as "sponge A" or "sponge B" and this will serve sufficiently well until specific identifications are made. In this way, the important fouling organisms and their seasonal settlement may be determined.

When the depth distribution is obtained, it may be possible to culture above or below this level. Particularly in the tropics, where the growth rates are rapid, it may be possible to utilize a growing season when fouling is minimal or absent. Knowledge of the fouling sequence and distribution opens the avenue for some options for avoidance where it is a serious problem.

PARASITES AND DISEASE

Disease in molluscs is probably the most difficult problem with which the culturist has to deal. Only in recent years has there been development of a reasonable body of knowledge on diseases of shellfish. The cells in molluscan body tissues are relatively small and difficult to study. There are seldom warning symptoms prior to the onset of a disease, and often the disease has almost run its course before it becomes apparent. Warning signs may be the appearance of "gapers" where the valves gape open but with the animal still alive or recently dead. This state does not persist for long, for scavengers, fish or crabs, soon remove the meats, and the shells are then termed "boxes" or "cluckers" from the empty noise when they are tapped. Another sign is the appearance of pustules on the surface of the body or pus in blood sinuses. Pustules, however, may only be a sign of stress rather than a disease symptom. Emaciation of the body, when normally the condition factor should be high, may be another indication of ill health.

Often mortality has occurred before there is an awareness something is amiss. The difficulty, then, is to determine whether the mortality was due to disease or to other chemical or hydrographic factors which may have persisted only briefly. Care must be taken not to expend too much time and effort on disease study. There are examples of decades of study on a single oyster disease without resolution of the cause. Also, it is rarely possible to actually cure the disease in large populations of molluscs in the marine environment.

In the unfortunate event of the occurrence of a disease, the objective should be to study and learn enough about its ecology so it may be possible to culture around it, thereby minimising its effect. On the Pacific Coast of Canada, such a situation exists with the Denman disease, for which even after 25 years the causative organism is only now being tentatively designated. It was found that the disease was most prevalent at the lower tidal levels in the intertidal zone and flourishes only during the months of April and May, presumably being influenced by water temperatures. By planting the beds after May and harvesting before April, or by using higher tidal levels, the main effects of the disease are evaded.

In the instance of *Perkinsus marinum*, a fungus disease of oysters in the Atlantic and Gulf of Mexico regions of the United States, the oysters are cultured to take advantage of the fact that mortalities and infections decline during colder periods of the year, when water temperatures are below 25° C, and when in salinities less than 15 ppt.

An example of one way in which disease control may develop is the case of the Malpeque disease in *Crassostrea virginica* in Eastern Canada. The disease, which has only recently been identified, began in 1915–16 in Malpeque Bay and by 1939 had spread to all the oyster areas in Prince Edward Island. Initial mortalities were nearly 100%, but by 1922 the few survivors in Malpeque Bay had reproduced well and formed the nucleus of a disease resistant stock. Oysters brought into Prince Edward Island invariably died. Attempts were made to keep the disease within the confines of Prince Edward Island and were successful for a number of years, but eventually it spread to the adjacent mainland. To enhance the speed of recovery of these stocks, disease resistant oysters from Prince Edward Island were introduced to provide an instant breeding stock and this procedure proved successful.

While there has been an awareness of some diseases such as Malpeque for a long time, most development in molluscan pathology has taken place in the last 20 years, owing mainly to disease outbreaks that have occurred during that period. The reason for some of these outbreaks is not clear, but most of them can be attributed to relaying indigenous species into a different environment or to the introduction of species from another country. Increased stresses from pollution no doubt contribute to the severity of epidemics.

The introduction of new species such as that of *Crassostrea gigas* from Japan to the Pacific Coast of North America in large quantities, soon after the turn of the century, without adequate precautions might be excused because of the ignorance of the possible consequences. A number of Japanese invertebrates were introduced with the oysters, several of which had deleterious consequences such as oyster drills, a nuisance seaweed, fouling organisms, a marine wood borer and possibly the so called microcell disease as well as a parasitic copepod. The positive results were the development of a modest oyster industry and a fishery for a Japanese species of clam accidentally introduced. In recent years, particularly with the advent of hatcheries and adequate means of quarantine, the safe, or nearly safe, introduction of exotics is possible. A more difficult problem, however, is the effect of the new species on the indigenous biota. This may mean competing with and possibly overriding and eliminating a native species, or on the other hand, its own elimination by susceptibility to native diseases.

The parasites of molluscs encompass a wide range of organisms from protozoa to molluscs themselves. There is also a wide range of effects varying from minimal such as food sharing to outright harm ending in mortality. Hence, the distinction between "parasites" and "disease" is not always clear, except that the organisms associated with disease are usually pathogenic. The following list indicates in a general way parasitic and disease–causing organisms that have been identified with molluscs. However, causality has not always been confirmed. It will be noted that practically all refer to the temperate zone, largely because this is where the study effort has been concentrated up to now.

Protozoa

Mastigophora. The commonest species is *Hexamita inflata*, a free–living flagellate, about 25 µm in length. It often occurs in dying oysters and is not considered a cause of death.

Sarcodinia. These are amoeboid protozoa and have been found in several oyster species but no pathogenicity has been established.

Sporozoa

Telasporea. One of the common species of this group is the gregarine, *Nematopsis*. There appears to be minor tissue damage from infection of this organism but it is not lethal.

Haplosporea

Minchinia nelsoni and *Minchinia costalis* are presumed to be responsible for extensive oyster mortalities on the Atlantic Coast of the United States in the 1960s. Another genus is *Marteila* with one species in France and one in Australia.

Microsporidae

Chytridiopsis mytilovum has been found in the ova of the mussel *Mytilus edulis* on the United States Atlantic coast, but it is not considered to be of economic importance.

Cileatea

There are a number of species, mainly of the genus *Ancistrum* which occur in a variety of molluscs, often on the gills. In the main, they are not considered to be disease inducing.

Bacteria

Species of *Vibrio* are found in oysters from the west coast of North America and in larvae in hatcheries on both coasts. A condition known as "focal necrosis" in *Crassostrea gigas* occurs in oysters from Japan and Washington State. "Bacillary necrosis" is another disease found in larvae in hatcheries from both coasts of the United States.

Virus

A herpes virus occurs in *Crassostrea virginica* haemocytes from Maine and another in *Ostrea edulis* in Great Britain. An iridovirus is found in the gill tissue in *C. angulata* and *C. gigas* in France, Portugal and Spain.

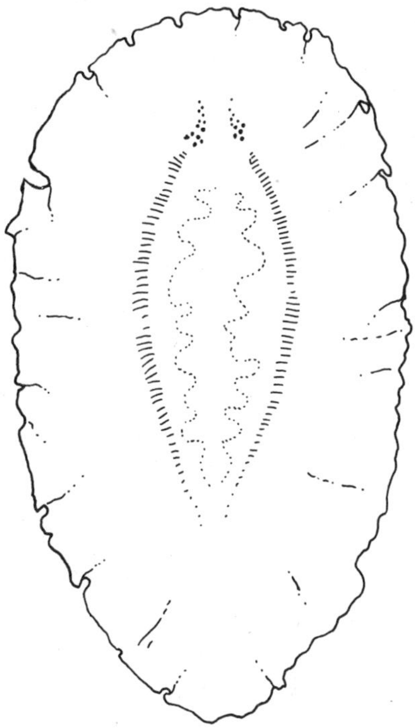

*Fig. 50. The oyster eating flatworm (*Pseudostyloclus ostreophagus*). x 10 (from Quayle, 1988).*

Fungi

A fungus–caused shell disease occurs in *O. edulis* in Europe. A mycelial disease (*Perkinsus*) has been found in *C. gigas*, *C. virginica* and *C. angulata* in the USA, France and the Netherlands.

Porifera

Although most sponges are epizooties, the genus *Cliona* is a boring sponge able to penetrate and honeycomb calcareous shells with fine tunnels. These may penetrate the nacre to the mollusc body proper, requiring the mollusc to secrete additional protective shell. The tunnels weaken the whole shell to such an extent it may easily be crushed and thus, cause shucking difficulties. The sponge, usually yellow in color, may be seen in the 1 mm diameter circular holes in the mollusc shell that give the sponge access to the external environment.

Platyhelminthes

Turbellarian flat worms of the genus *Stylochus*, *Pseudostylochus* and others known as oyster leeches are probably more predators than parasites. Most *Stylochus* are probably post mortem attackers, although they are said to be able to attack young spat. *Pseudostylochus ostreophagus* (Fig. 50), a Japanese species introduced to the west coast of North America with oyster seed, is able to drill an oval perforation in the shell of young oyster spat up to 1 cm in diameter and is capable of causing considerable mortality. Oyster leeches have been found within the shell of living oysters with some evidence of active feeding.

Trematoda

Numerous species of this group infest a wide variety of molluscs. The life history is quite complicated, consisting of various stages, some of which occur in other species such as fish or birds. The effect on host molluscs is mainly on metabolism or on reproductive capability. They are seldom lethal. Well known genera include *Bucephalus* and *Gymnophallus* with circum-boreal distribution in a number of molluscan species. The size ranges from 0.25 to 0.5 mm in length.

Cestoidea

Molluscs serve as intermediate hosts for this group which includes *Echeneibothrium* and *Tylocephalum*, the latter known to occur in the tropics. *Tylocephalum* can be about 1.5 mm in length. *Echeneibothrium* has been found in clams.

Nemertinea

The common nemertean occurring in close association with lamellibranchs, usually in the mantle cavities, is *Malacobdella*. They are considered to be commensal rather than parasitic. The free–living *Cerebratulus* may be a predator on some molluscs.

Nemotoda

Several nematodes have been recorded from molluscs, but their role is uncertain except in the case of *Echinocephalus*. This species occurs in the foot of the abalone (*Haliotis*) and weakens clinging ability. *E. sinensis* infects the genital duct of *C. gigas* resulting in tissue damage. (It may also be a human health hazard.)

Annelida

Polydora is the polychaete most closely associated with oysters. It drills into the shell edge and when it penetrates the nacre it may induce additional shell deposition. The problem in the temperate regions is the unappetizing appearance of half–shell oysters. In some instances, mortalities have occurred and it can be a serious pest.

Mollusca

The pyramidellid gastropods, which are ectoparasites, are mainly in the genus *Odostomia*. They are small, less than 1 cm in length, and attach to the shell edge of lamellibranchs and feed on host tissue fluids by inserting a long proboscis into the body. Others live within the mantle cavity. *Odostomia* may interfere with the mollusc's development and growth and cause shell deformation, particularly in young oysters.

Crustacea

There are a number of copepods parasitic in molluscs but probably the most important is *Mytilicola*, sometimes termed the "red worm" (Fig. 51). This is an elongate crustacean between 5 and 10 mm in length which lives in the intestine and rectum of lamellibranchs and gastropods, sometimes numbering up to six specimens. *Mytilicola* is known to have caused mortalities, particularly in mussels, and can seriously affect the condition of mussels and oysters. Other genera, such as *Ostrincola* and *Myicola*, live in the mantle cavity of other molluscs.

Pinnotherid crabs live within the mantle cavity of bivalves. In most cases they cause no great harm and are labelled as symbionts. However, particularly in oysters there can be ctenidial (gill) damage and reduction in

*Fig. 51. The parasitic copepod (*Mytilicola orientalis*). x22 (from Quayle, 1988).*

condition. It may be possible that the presence of pinnotherids may affect the sex ratio in oysters if there is interference in food supply, as postulated for the tropical oyster *Saccostrea cucullata.*

Disease Summary

1. Mortalities may be caused by disease, predators or physical and chemical disturbances. If there is a mortality or if the species looks or reacts abnormally it is not necessary to rush into a detailed pathological study.

2. If the causative organism is apparent, have it identified by a specialist in that group.

3. If there is no apparent organism, examine the physical and chemical status of the immediate environment.

4. Study how the situation changes with season, with site, and alterations in the hydrographic factors.

5. Culture to minimize the effects of the disease.

6. If there is significant mortality of unknown cause, concentrate the survivors to create a possible nucleus of a disease resistant population.

7. Keep careful records.

PREDATORS

The hard shell of most molluscs and the ability of many to close up tightly makes them immune to many, but not all, predators. Some predators are specially designed to penetrate shell in one way or another. To assess actual and potential predator damage that may occur in a culture situation, estimates of predator abundance and a knowledge of their life history is necessary. This is important for it may show one or more weak links that can be attacked to render the predator either impotent or at least reduce its effect.

Gastropods

Certain snails are ubiquitous bivalve predators. They are known as drills, from the ability to pierce bivalve shells. This is done by means of an extensible, flexible proboscis that can extend well beyond the mouth aperture. Attached to the proboscis is a radula, a ribbon studded with horny teeth, which passes back and forth as on a pulley, and in this way creates a rasping action. Associated with this apparatus is an accessory boring organ which is a soft pad. When applied to the shell of a prey the secretion of the boring organ softens the shell enough to allow the rasping action of the radula to scrape out the shell. By repetition of this process, the shell is penetrated to give the radula access to the flesh of the mollusc victim.

Two gastropod families most often involved in bivalve predation are the families Tritonidae and Naticidae (Figs. 52 and 53). Several species of tritons are oyster drills. The naticids, known as moon snails, are associated with clam predation. These snails are burrowers and able to attack their prey under the surface, holding them in place during the drilling action with a large muscular foot.

Gastropod predators, if not too abundant, may often be controlled simply by removal by hand at frequent intervals. If they are very abundant it may be necessary to try control by trapping in the immediate vicinity of racks or bottom culture operations. The traps are simply small cages of chicken wire or fish net baited either with molluscs (clam or oysters) or fish remains after filleting. These control practices may not completely eliminate the drills but may reduce the population to a level where the problem is not too serious. Naticid drills are difficult to control except by picking. Careful records should be kept to determine effectiveness of control measures.

Since most predatory tritons do not have a pelagic larva, suspended culture operations are generally immune to attack by snails, so that only bottom and rack culture are in potential danger. Some snails, and this is typical of many Tritoniidae, congregate for breeding and egg deposition and this is an appropriate time to collect them. Raised objects such as boulders, which serve as capsule–depositing sites, attract snails.

Crabs

Both bottom and suspended culture are open to predation by crabs, the former by bottom dwellers and the latter by swimming crabs. Small molluscs down to a few mm are preyed upon by crabs, but cancroid as well as portunid species with their strong pincers are able to crush quite large shells.

Control of crabs is by picking or trapping. Traps may be made in several ways (Fig. 54). A typical form is triangular in shape, made of laths, sticks, or webbing with an opening at the top. Baited ring nets, simply a ring of metal or wood about 1 m in diameter and covered with a fishnet of a mesh size small enough to catch quite small crabs, may also be used. The main feature of this trap is that it must be hauled to the surface fairly frequently, at least before all the bait is consumed. Another method is the trot line, a form of set line, with baited snoods, not necessarily with hooks. The crabs continue to hold onto the bait when the line is hauled, so they may be collected with a dip net just below the surface. Here again, the trot line must be hauled before all the bait disappears.

Careful records are required to assess the effectiveness of crab removal. As with drills it may not be possible to remove all of the crabs, but the level of predation may be reduced.

*Fig. 52. a. Hole drilled by the moon snail (*Polinices lewisii*) (x 0.5). b. Moon snail (x 0.5) and egg case (x 0.75) (from Quayle, 1988).*

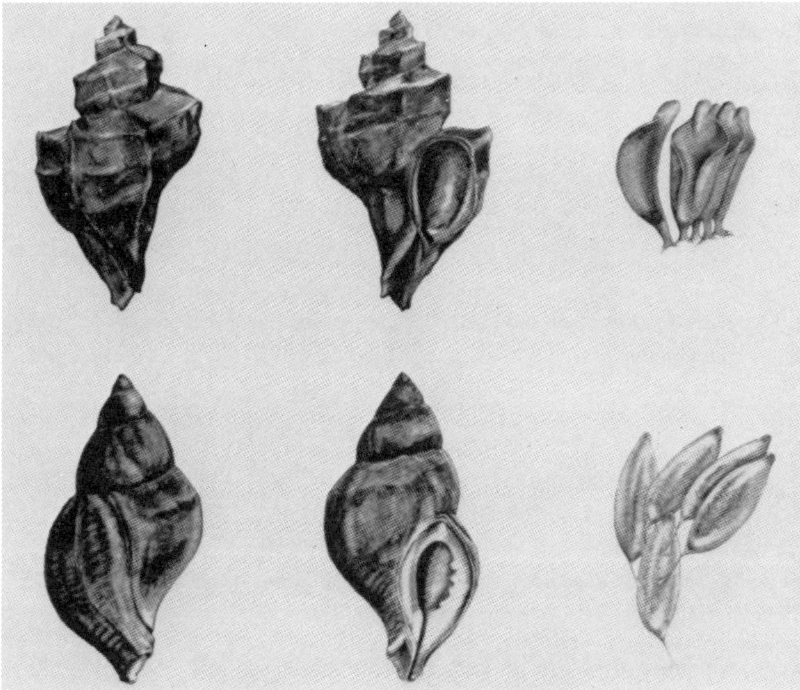

Fig. 53. Comparison of the shells (x1.3) and egg capsules (x 3.6) of the Japanese oyster drill (top) and those of the native drill (bottom) (from Quayle, 1988).

Seastars (Starfish)

Seastars may be important predators particularly with bottom culture, but they may also be a factor in suspended culture for they have pelagic larvae and this allows them to settle on material off the bottom. In intertidal culture, they may be readily removed or destroyed by dropping a teaspoon of carbide or quicklime on them. It is nearly impossible to destroy a seastar by cutting or mutilation since they will regenerate lost parts.

To open a bivalve, the sea star clasps it with its suction tipped tube feet with which it exerts pressure against the adductor muscle which is then torn by the force. The seastar also has an evertible stomach that is able to penetrate

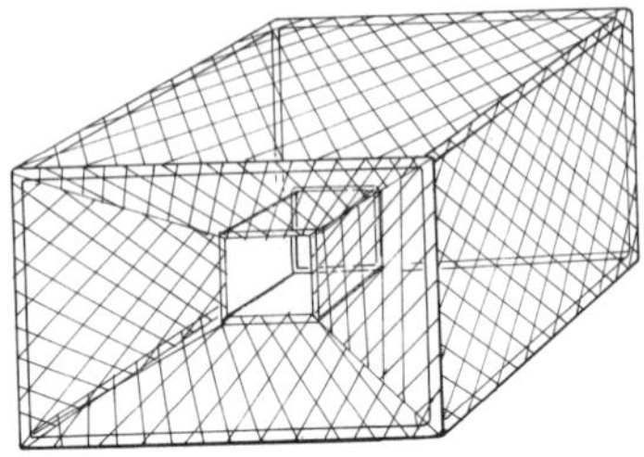

Fig. 54. Simple crab traps. Top — wire mesh. Bottom — wooden lath. (x0.125) (from Quayle, 1988).

very small apertures. Preliminary digestion of the mollusc body takes place, and this eventually causes relaxation of the adductor muscle causing the valves to gape and allowing full entry.

An oyster 75 to 100 mm in length may be devoured by a seastar in less than 24 h. Some species of seastars are able to dig with the rays and tube feet to reach burrowed clams. Knowledge of the breeding season, growth rate and abundance is useful in combatting whatever danger they impose. Young seastars tend to evade light so they are often not readily visible.

Seastars are serious problems for subtidal bottom culture in some parts of the world. Spreading quicklime on the bottom destroys seastars but does not harm the molluscs. Vessels towing an array of large string mops over the bottom is also another system of control. The spines and pedicillaria of the seastars become entangled in the strings of the mops which, when lifted to the surface, are dipped in vats of hot water, killing the seastars.

Up to now seastars do not appear as much a problem in tropical bivalve culture as in temperate waters. However, a devastation of corals caused by the seastar *Acanthaster* in tropical waters suggests the possibility of difficulties, although the corals and *Acanthaster* do not occur in estuaries where oysters are found.

CHAPTER 5

GENERAL ASPECTS OF BIVALVE CULTURE

The oldest known form of bivalve culture is oyster culture which dates back to the time of the Romans who prized oysters highly. It is said that one of the reasons for conquering and holding the British Isles was because of the abundance and quality of the oysters there (*Ostrea edulis*). Temperate zone oysters have been cultured for many years. Established techniques for culturing oysters have been used in western Europe, North America, Australia and Japan. The books by Korringa (1976 a, b, c) give reviews and further references. The culture of other species has developed more slowly, but there has been a rapid expansion of the culture of mussels, in particular, and other species in the past 20 years.

In spite of large natural bivalve populations in many tropical areas, culture has not developed very much. There have been many attempts at tropical oyster culture but with few lasting results. Noteable exceptions are in the Philippines and Thailand. The culture of blood cockles in Malaysia has developed rapidly in the past 20 years.

Development of culture systems in temperate waters has been a long slow process, still incomplete. Laboratories, dedicated almost entirely to oyster biology and culture, have been in existence almost since the turn of the century and are still functioning. As a consequence, there is extensive literature on oysters and oyster culture. It is said there are more publications on oysters than on any other invertebrate group, but these are concerned mainly with temperate species. However, the basic principles of biology and culture so developed are, to a large extent, applicable to the tropical situation. Rapid growth, long breeding season, and salinity as a dominating factor in the tropics rather than temperature are major differences from temperate conditions.

Actual production of bivalves is only a part of the development of a whole industry. When animals are cultured rather than harvested from the wild, they become property. As with farm produce, control or ownership of the land on which it is grown is necessary, and fish and shellfish are no different. This requires a mechanism such as a leasing authority to administer and control intertidal or deep water leases, or in some cases the sale of such property. Another important phase is that of sanitation control which requires bacteriologists and trained shellfish sanitarians. These are but two examples of the infrastructure required for a bivalve culture industry.

In this section some of the general aspects of bivalve culture will be discussed. These include the important first step of selecting a culture site. A number of considerations go into this process. As part of the process of site selection and for ongoing studies it is necessary to study the local oceanographic conditions. A discussion is presented of the relevant parameters. Actually getting out into the field to do the studies requires the ability to find directions, locations and read charts. The basics of these skills are discussed.

CHOOSING A SITE

Choosing a site for bivalve culture is the first, and perhaps the most important, step. The obvious consideration in evaluating potential sites is whether the animals will grow well. This is a biological question that takes into consideration a number of ecological parameters. However, good growth is not the only requirement for successful bivalve culture. A number of other factors that affect management and the economics of culture must be considered. These factors range from the effect of physical parameters to political forces. The topics of concern are:

1. Water depth
2. Bottom characteristics
3. Protection from wave action
4. Water quality
5. Tidal flow and height
6. Turbidity
7. Predation
8. Fouling
9. Pollution
10. Navigable waters

11. Access
12. Conflict of uses
13. Permits

The species and the culture system must be suitable to the site. In choosing a site, consideration should be given to the possible species and culture system combinations that might be used. The "match" of the site characteristics with the species and culture system has to be evaluated.

The importance of careful site selection cannot be overemphasized. There have beeen many cases of poor site selection in aquaculture. A site might be "acceptable " or "reasonable" but, with care and thorough evalu- tion, better sites might be found. Unfortunately, what often happens is an investment is made in a site, for example shore–based facilities, that makes it difficult to change sites later.

Some of the problems of site selection arise from inadequate attention paid to the factors discussed below. Other considerations are often given greater value, like previous ownership of shore front property, or ease of access in initial stages due to proximity of residences. These factors should certainly be considered but the short–term benefits and costs have to be weighed against the long–term profitability of a site.

The most satisfactory approach to selecting a site is to carry out experimental culture at a number of potential sites identified by a regional survey. A minimum of one year is recommended to assess the seasonal variation. Larger or smaller experimental units may be used based on the information gathered in the site survey. Scaling up will take place as confidence is gained in particular sites.

The topics will be discussed in general here. The particulars as they apply to certain species or methods of culture will be discussed in subsequent sections.

Water Depth and Bottom Characteristics

The species and the culture method will determine what type of site is suitable as far as water depth and bottom characteristics are concerned. For bottom culture, the bottom consistency must be such that it will support the seeded cultch or adult animals or, for clams, be appropriate for the species. For most forms of suspended culture (racks, longlines, rafts), the type of bottom is of little concern. However, even with floating systems, bottom consistency will affect the kind of anchoring needed.

Water depth will modify the details of various culture methods. For hanging culture methods (racks or floating), the depth must be adequate for the length of strings to be used. Very deep water means that longer anchor lines are needed for floating structures. These aspects will be considered as each culture system is discussed in subsequent chapters.

Tidal Height

Tidal range varies extremely from one part of the world to another, even within short distances. In addition, the nature of the tidal oscillation varies. In a diurnal tide there is only one high and one low tide per day. In a semidiurnal tide there are two complete oscillations with two highs and two lows each day.

Tidal height (or range) is important in evaluating the type of culture system to be used. If the height is small there is little exposed intertidal area. A small range will also affect the management of rack culture, since the exposure to the sun and air is often necessary for fouling control. Tidal height is of little concern in floating systems except as it affects the anchoring system.

Tidal Flow

Tidal flow is of concern mainly where there is a large tidal range at or near narrow passages where flows up to 10 or more knots may occur. There may be movement of strings similar to that caused by wave action. In general, tidal flow is of greater concern in suspended culture than in bottom culture because of string or tray movement from current action.

Protection from Wave Action

Regardless of the culture system, protection from waves is required. Strong waves will stir up sediments in shallow water reducing feeding efficiency. Extra anchorage of floating systems may be needed. Wave action will cause jerking movements of suspended ropes and trays enough to cause damage.

Sites within estuaries, bays or lagoons are preferable. The area of the water surrounding the site will have some influence on the potential waves that can develop. The larger the area over which the wind blows (the fetch), the higher the resulting waves will be. The wind speed and length of time it blows influence the height of waves. Since wind direction and force vary seasonally, the annual cycle must be considered. Local fishermen will be able to provide first estimates of the amount and seasonality of rough seas.

In some cases it may be possible to avoid or minimize the effect of seasonal increases in wave action. In the tropics it may be possible to complete a culture cycle between the seasons of high seas. This has been proposed as a solution to mussel culture in southern India where the monsoon season makes raft culture impossible during the whole year.

Water Quality

Water quality includes temperature, salinity and chemical constituents such as nutrient salts and food. Most of these are fixed factors for any one site and little can be done to alter their values. However, advantage may be taken of slight variations within short distances. If there is already a population of bivalves, it would indicate that the water quality is reasonably suitable. If an introduction is being considered, either of a native or an exotic species, then the temperature and salinity should meet the requirements of that species. The broad ranges of suitable levels of temperature and salinity are discussed in the sections for each species group.

There is often undue emphasis, particularly in relation to new projects, on nutrient and food studies. However desirable they may be, they are most difficult to conduct, requiring a time–consuming, long–term effort with costly equipment and well trained personnel. As with other water quality data, the best that such studies can provide is a possible explanation for events. The bivalve itself, by its own reaction to the environment, provides a measure of water quality, for it integrates all factors as reflected in shell growth and condition of the body.

Temperature and salinity should certainly be recorded, for in the tropics salinity is the main variable. Hydrogen ion concentration measurements are of little value except in special physiological studies or perhaps in occurrences of excessive flooding. There are many tasks to be done without spending valuable time recording measurements of questionable value.. Other factors that affect water quality will be considered under the topic of pollution.

Turbidity

Unless turbidity is at a level great enough to cause silt deposition sufficient to cover oysters on the bottom, it is not of great concern. Indeed, many of the great oyster producing areas of the world occur in areas of relatively high turbidity. It is not known where the dividing line may be between a level of silting that causes problems of excessive energy consumption during filtration and that level that may contribute to bivalve nutrition. The level of siltation tolerated by large bivalves may smother seed, so consideration of the phase of culture is important. In some instances, it may be necessary to study the rate of silt deposition in turbid areas if there are indications of deleterious effects. This involves comparative examination of particle size and the amounts collected in sediment traps at various sites.

Pollution

Two types of pollution of prime importance to bivalve culture are domestic sewage and industrial sewage. Domestic sewage pollution may originate from:

1. Discharge of untreated or partially treated sewage from trunk sewers
2. Discharge from malfunctioning or improperly installed septic tanks
3. Land runoff through seepage after rain

4. Rivers and streams
5. Shipping or boats.

Problems of dealing with bivalves from contaminated areas, called shellfish sanitation, are dealt with in Chapter 10.

Industrial pollution emanates from industrial activities like pulp milling, logging or mining. Pulp mills require large quantities of water, and their discharge carries waste liquor containing a variety of chemicals, some of which are inimical to the well–being of shellfish.

The obvious place to discharge this waste water is in bodies of water such as rivers, lakes, or the sea. Mills attempt to recover as much material in waste water as possible for economic reasons but for a similar economic reason they do not extract all of it, so some chemicals and fibre escape. An average sized pulp mill may discharge more than 100 million litres a day, equivalent to a small stream. These effluents, sulphite waste, liquor or Kraft mill effluent, when discharged into salt or fresh water, may affect organisms living there by toxicity, increased oxygen demand, and increased particulate material.

Toxicity

The source of toxicity is the complex chemical constituents of the waste. There may be a direct toxic action on the shellfish itself or an indirect effect on the food organisms on which the shellfish rely.

Oxygen demand

The decomposition of organic materials in the effluent requires large quantities of oxygen. This requirement, called biological oxygen demand (B.O.D.), may deplete the surrounding waters of oxygen upon which most living organisms depend. Excessive B.O.D. may cause suffocation of organisms living in the area. Mobile species like fish can escape but not sessile invertebrates such as bivalve molluscs.

Particulate material

With pulp mills this consists of wood fibre, bark and chips, the heavier particles of which sink and form a mat on the bottom. This prevents water circulation. Materials under the mat decompose, creating an oxygen demand and mortality in the fauna living there. This mat may also prevent further settlement of invertebrates, and hydrogen sulphide bubbles may eventually be seen bubbling up in the area. Lighter suspended material, if in sufficient quantity, may clog the gills of filter feeders such as bivalve molluscs.

All these factors lead to a general deterioration in the abundance, diversity and well being of the organisms normally living in the area. Oysters and other molluscs may be affected, resulting in:

1. Mortality
2. Reduced growth rate
3. Reduced fatness or
4. Reduced breeding success.

Massive mortality is clear cut and unequivocal. However, mortality, growth rates, fatness and breeding are widely variable in non-polluted areas, and unless pollution effects are quite drastic, the response of the bivalves in the polluted area may fall within the usual range of variability. An annual average of monthly determinations of condition factor will take into account normal seasonal variation of bivalves owing to the sexual cycle. For example, the usual range of the condition factor index in oysters is between a low of 50 and a high of 150. A one year study in a polluted area may produce an annual average of 80. The question is whether this low level is caused by pollution or simply a poor year. The only way this may be answered is by comparing results for several years of study with those from adjacent growing areas where the possibility of pollution is non-existent. These must be carefully designed experiments with enough replicates to determine normal variability at each site so there may be adequate statistical comparisons.

Pollution studies have often been taken into the laboratory, measuring mortality, growth, oxygen consumption, heart beat, etc. against varying concentrations of effluent. A difficulty lies in applying these results from set values of various parameters to the field, where these are in a continual state of flux. Care must be taken, as with disease, not to divert too much time and effort to the problem. Attempting to culture bivalves in polluted areas should be avoided if possible and effort expended to prevent pollution in established growing areas.

Pulp mills are only one of many forms of industrial pollution. Tailings from mines or smelters can contaminate either by leaching from shore dumps or actual deposits in the sea. A typical and readily apparent example is copper mining where leachates are able to colour the meats of oysters growing in the area a violent green, rendering them unmarketable. However, much lower levels of heavy metals can debilitate bivalves and be a human health hazard.

In areas where logging is carried out, logs are usually dumped, collected and stored in protected bays which are often potential or actual oyster growing areas. Bark debris can destroy the bottom and, if the water is shallow, propellor wash of tug boats can create furrows on the bottom; if oysters are present, they are washed into windrows. Bivalve culture and industry in one area are seldom compatible.

Tributlytin antifouling paint is causing concern where many vessels occur in the neighborhood of bivalve culture. Antifouling paints operate on the principle of leaching the toxic elements from the paint coating to form a thin protective layer of a poisonous solution. Turbulence disturbs this layer and causes distribution of the toxic solution beyond the object being protected.

Both laboratory and field studies have shown that organotin compounds such as tributyltin oxide in antifouling paints affect shell deposition in Pacific oysters. It leads to a thickening and layering (chambering), particularly of the upper (right) valve, sometimes forming inclusions of gelatinous material between the layers. Growth and condition factor are reduced.

A high correlation between these phenomena in the oyster shells and the proximity of vessels has been esablished. In some places like France the use of antifouling paints containing organotin has been banned on vessels less than 25 m.

Predation and Fouling

The species that may cause problems of predation and fouling were discussed in Chapter 4. It may be difficult to assess a site with respect to potential predators or fouling species in a cursory survey. This will probably require a longer term study with experimental culture. However, a preliminary survey may indicate potential problems. A large barnacle population on intertidal rocks may indicate a potential problem for spat collection or growing. Thick growth of macroalgae in low intertidal or subtidal areas, or growth of algae and invertebrates on pilings or mooring ropes, suggest potential fouling problems. However, these should not necessarily be reasons for rejecting a site, as fouling or predation may be within manageable limits. Indeed, healthy populations of such organisms indicate high productivity which should also mean good bivalve growth!

Navigable Waters

Many countries have legislation or regulations governing navigable waters. The jurisdiction may be with the Coast Guard, municipalities or other bodies. Placing floating structures in waterways (such as longlines or rafts) often interferes with boating and, thus, is subject to control. If such control is present in the area of concern, then official permits may be required.

Aside from the official control of waterways, there is the "law of the sea" which will be an unwritten law often with local variations. Violation of locally accepted practices and territoriality can lead to serious consequences. The introduction of bivalve culture operations in an area where there has been fishing and boating for many years may lead to conflict. Setting up longlines, rafts or racks can interfere with traditional fishing grounds.

It will be true in many places that the addition of the culture structures (racks, rafts, etc.) will increase fish populations since they act as artificial reefs. However, it will be difficult to convince fishermen of this in advance. They will see the structures as interference. Regardless of what legal rights one has to an area, if the local residents object there are bound to be problems.

Access

Some of the best culture sites may be very difficult to reach by land or sea. If the cost of transportation is going to be excessive, it will preclude an economically successful venture. Whether access to a site from either the shore or by sea is necessary will have to be determined. It may be that one or the other is sufficient. Access is needed to transport in culture materials and to carry out the product. If access is to be by sea, consideration must be given to seasonal weather and sea conditions.

Conflict of Uses

As indicated above, there are potential conflicts with other users of potential culture sites. Some multiple use is possible without conflict. Some conflicts can be resolved. In most cases, the culturist is the new arrival and in the least favorable bargaining position; unless it can be argued that this new enterprise will be a benefit to the local community. Since most culture operations are exposed to vandalism, it is essential that conflicts be avoided or resolved with the local residents. Official permission will not protect the investigator or the investor from vandalism.

During the site selection survey, the pattern of local fishermen and other users of the water should be determined. It may be possible to adjust the location or culture method to minimize conflict. Discussions with key fishermen or residents will often help to identify potential conflicts. An open discussion of the development plans, indicating the benefits to the residents, will often prevent suspicions which create rumors.

Permits

If marine aquaculture is new in the region, there are probably no regulations controlling its development or permits available specifically giving the culturist legal access to a site. However, if a culture industry is going to develop (artinsinal or otherwise) it is necesary that it be given some legal status. As an example, until legislation specifically provided for it, legal ownership of stock in suspended aquaculture was not available in some provinces of Canada. It was possible to obtain permission to place the structures in the sea, but the fish or shellfish were not "owned" by the farmer until they were harvested!

There may be existing laws or regulations that apply to bivalve culture, even if they were not originally enacted for that purpose. These should be determined for the local situation. It would be preferable to have regulations that give the farmer permission to operate within a specific area, with the areas allowed being large enough to allow a viable operation, and with requirements for some kind of performance standards to maintain productivity.

OCEANOGRAPHY

Oceanographic factors of most concern to bivalve culturists are temperature, salinity, tide and currents. Of lesser importance are hydrogen ion concentration (pH), oxygen content and turbidity. Other chemical constituents, such as nutrient salts (nitrogen compounds, phosphates and silicates), can influence production of potential bivalve food but are difficult to measure. Furthermore, it is almost impossible to integrate nutrient chemistry data in terms of potential bivalve production.

Anyone growing bivalves will have to know the oceanographic characteristics of the culture site and the seasonal variation. Those involved in research to develop or evaluate culture in a new area should keep careful records of a number of factors in order to understand changes in the animals and the differences between sites. If there are existing records of nearby sites, even if they are from deeper water, they should be examined. Data should be collected on a regular basis, and simple measurements like temperature and salinity should be done at least weekly. Depending on the tidal situation, it may be important to determine the daily variation.

Data should be kept in record books but graphs should be maintained to which new data are continually added. Graphs will give a much clearer presentation of the information and allow easy comparison to information on condition factor, spatfall, growth, etc. Often data, such as the oceanographic data, are recorded in log books and not examined until it is time to write a

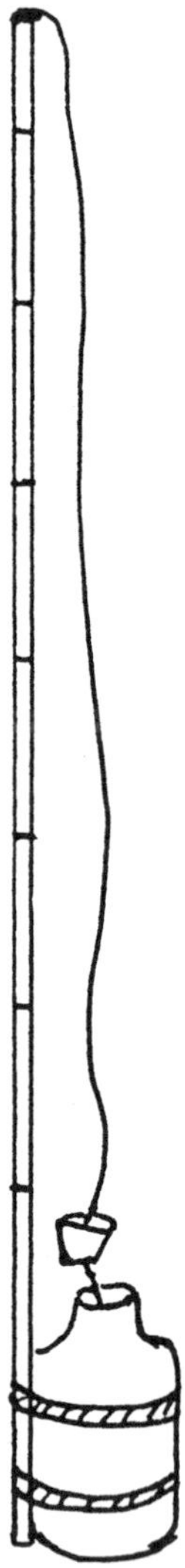

Fig. 55. A water sampling bottle fixed to a stick for working in shallow waters.

report. However, changes that take place in the field on a weekly, or even a daily, basis are important for management decisions. It is recommended that graphs be kept on the office wall with a continuous, up-to-date record of the data.

The emphasis in this book will be on simple methods and equipment for fieldwork. Stem thermometers and refractometers are recommended instead of electronic devises. The electronic equipment is great when it works and when it is calibrated. However, under field conditions it is sometimes difficult to handle and be sure the calibration is maintained. The electronics do not necessarily give a more accurate measurement of the parameters. This depends on the sensitivity of the instrument which is usually a function of the cost. Since in most situations readings to the nearest degree, as an example, are as accurate as is needed, the added expense of purchasing and maintaining instruments is not warrented.

Temperature

The simplest way to measure surface temperatures is to take a bucket of water into the boat, immerse a thermometer in it and wait until movement of the mercury column stabilizes and take a reading. Standard glass thermometers should be protected by a metal container. The high water slack of the day is an appropriate time if only one reading is taken. In a similar, way subsurface temperatures can be recorded by taking a water sample with a bottle. If the temperature is recorded immediately, there will be little loss of accuracy. A simple water sampling bottle is shown in Fig. 55. When the bottle is at the required depth the stopper is removed with the pull on the string. When the air bubbles have stopped the bottle will be full. The stick can be replaced by a rope for deeper water, but in this case the bottle must be weighted so it will sink.

In situations where temperature varies considerably, it may be worthwhile to have a submersible, continuous temperature recorder. Each unit is expensive but will provide a continuous record showing all temperature fluctuations within the period at that depth. This permits an analysis of the temperture (i.e., degree-days) to which the bivalves in the area are subjected.

Another useful instrument is the maximum–minimum thermometer, which records the maximum and minumim temperature for the observation period. These are inexpensive. The observation period can be recorded daily or weekly but must be done manually.

Salinity

Salinity is a measure of the salt content of seawater. Chemical measurements of salinity are based largely on the determination of the chloride ion concentrations rather than on sodium chloride. Both salinity and chlorinity are expressed in g/kg of seawater or in parts per thousand or per mille. The symbols $^{o}/oo$ or ppt are used.

Measurements of salinity are most easily made by using the physical properties of seawater which vary with salinity. The two easiest methods are use of a hydrometer and a refractometer. A hydrometer actually measures the density of seawater which for a given temperature is largely determined by the salinity. Hydrometers are accurate enough for most shellfish work but are fragile. The refractometer is a sturdy, handheld instrument and requires only a small water sample. It measures the refractive index of the water which, like density, is a function of the salinity at any given temperature. Refractometers are moderately expensive but with care are long–lasting.

Note that salinity measurements by density and refractive index are influenced by temperature. This is the temperature at the time of measurement. The nice feature about salinity is that it need not be measured in the boat on site. A water sample can be taken ashore and stored without measurable changes in salinity as long as there is no evaporation. This also saves field time. It is recommended that the hydrometer or refractometer not be taken out in the boat. This reduces the risk of loss. The sample should be measured on shore or in the laboratory. It will also be easier to obtain an accurate reading.

Oxygen

The oxygen content of seawater in which bivalves will be grown is important. In most natural circumstances there are adequate amounts of oxygen. Extraordinary circumstances include reductions caused by biological oxygen demand from pollutants or from decaying organisms such as plankton blooms and certain upwelling circumstances.

If bivalves live naturally in an area, it may be assumed that the oxygen supply is satisfactory. Temporary oxygen depletion for a number of days usually has no significant effects on oysters, as they close their valves and live for a time without an external source of oxygen. Other bivalves may be more sensitive.

There are electronic instruments for measuring the oxygen content of seawwater, or chemical methods can be used (the Winkler method). Unless there is good reason to suspect problems, the routine measuring of oxygen is not considered a necessary part of site evaluation.

Hydrogen Ion Concentration (pH)

As with oxygen, if shellfish are living in an area it can be assumed that the pH is suitable. Changes in pH may be caused by reduced salinity, but fairly drastic differences over a period of time are necessary to cause difficulties with bivalves. Unless abnormal conditions develop it is unnecessary to be concerned about pH in bivalve culture operations.

Hydrogen ion concentration is measured with electronic instruments (pH meters) or colorimetrically. In the main colorimetric technique, a controlled amount of indicator solution such as cresol red or phenol red is added to the sea water sample and the color developed is compared to that of a set of standardized tubes. This method is accurate enough for most shellfish field work.

Turbidity

Since many bivalve areas are in estuaries, turbidity (transparency of water) exists in varying degrees and has a direct influence on bivalve culture. Turbidity may be caused by the silt load, by detritus (suspended organic material), by plankton or by a combination of all three. Turbidity may result in the deposition of silt which, in sufficient concentrations, may smother bottom living organisms. It also affects the feeding efficiency of bivalves when in high concentrations, since energy is expended separating out food and discharging unwanted particles. Different species will have different tolerances for high turbidity.

Turbidity may be measured by testing the limits of visibility or by light transmission measurements. For most studies the limit of visibility is adequately and simply measured by the Secchi disc. This is a circular plate, usually of metal, 20 cm in diameter with a line tied through the center. It may be all white or divided into four quadrants, alternately black and white on the upper surface with the lower surface black to prevent reflection of light. The Secchi disc is lowered into the water on a measured line. The depth at which the disc is no longer seen is noted as well as the depth at which it reappears when lifted. The mean of the two readings is the limit of visibility and is a rough but useful measure of turbidity. The time of day, amount of cloud cover and wave action affect readings, so these should be noted with each reading. Readings may be standardized at midday on the shaded side of the boat and preferably with a water glass and observations made at 1 m above the water surface.

Tides

Tides are an important factor in determining the type of culture in a given area. Tides are generated mainly by the gravitational effect of the moon and sun, particularly the former. Wind and barometric pressure will exert a smaller influence in certain localities. For most areas in the world, it is possible to predict fairly accurately the movement of tides. The predicted times and heights are published for most of the shipping ports. A typical tide table is shown in Table 4.

Day	Time	Ht (ft)	Ht (m)
1	0135	14.8	4.5
	0850	5.0	1.5
	1535	12.5	3.8
	2020	9.5	2.9
2	0220	14.5	4.4
	0925	4.2	1.3
	1630	13.3	4.1
	2120	10.0	3.0
3	0250	14.2	4.3
	1005	3.5	1.1
	1720	13.9	4.2
	2225	10.3	3.1

Table 4. Typical time and tide chart.

In the table it can be seen that on June 1 at 0135 the height of the tide is 4.5 m above chart datum, which is the plane below which the tide seldom falls and is the basis for making the charts. Approximately 6 h later at 0850 the tide has fallen to 1.5 m above chart datum. There is then a succeeding high and low tide. Thus, there are two complete tidal cycles or oscillations per day, and this is termed a semidiurnal tide. If there is only one complete cycle in 24 h it is called a diurnal tide. In some areas there are no tides or a range of only a few centimetres.

On approximately a 28–day cycle, the relative positions of the sun and moon change. When they form a straight line with the earth, that is at the full moon and the new moon, they pull on the tides together. This causes a greater range in the tide: higher high tides and lower low tides. These extreme tides are called spring tides. Two weeks later, in between the spring tides, the sun and moon pull at right angle and partially cancel each other's effect. Thus, the tidal range is least. These are called neap tides. The term spring tide has nothing to do with the spring season. The range of the tide changes with the seasons as the earth moves closer or farther away from the sun.

In addition to the tidal data in the published tables for the reference port, there is information on the time and height corrections to be applied for other ports. If the culture area is far from a secondary port it may be necessary to determine the appropriate correction factors. This can be done by setting out a graduated pole to record the tide levels. By making observations, particularly trying to determine "slack water," the time when the tide is at a maximum high or maximum low and is thus changing direction, a rough indication of the time and height differences can be determined. It may take a month or more of observation to find the corrections.

Currents

Currents are water movements developed by differences in tidal levels or induced by wind. In open waters, currents generally have small velocities, in the range of 1 or 2 km/h. In constricted areas such as estuaries or island complexes, currents may be quite rapid and speeds of 25 km/h are known. Currents are important to bivalve culture as they affect the position of beds, and the stability of structures such as racks or rafts. They also are responsible for the distribution of bivalve food (suspended particles) and larvae. Current speeds and directions should be investigated over several tidal cycles and, if possible, at different depths.

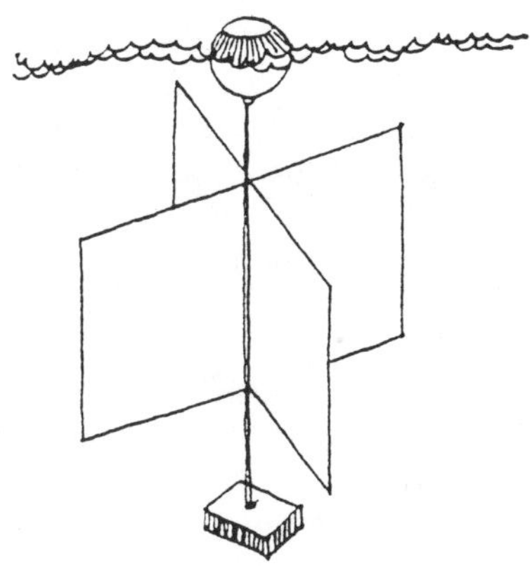

Fig. 56. Current drogue (from Quayle, 1980).

There is a wide range of complicated mechanical current meters but such meters are not necessary for most shellfish work. Drift poles, which are simply lengths of wood or poles (bamboo) weighted at one end so they float upright, have been used extensively. Depth may be varied with the length and weight used. For surface currents, plastic envelopes or other simple surface floating materials (even oranges) are satisfactory. Also in general use are drogues which are simply four plane surfaces of wood (plywood) set at right angles to a common centre (Fig. 56) and weighted to float at a certain depth. Either one or several may be set out at one or several stations and at different times to cover varying weather and tidal conditions. From direct observations and time, rough current patterns may be developed. For more accuracy, some means of taking bearings such as sextants or a bearing compass with or without range finders would be required.

MARINE CHARTS

In bivalve culture, some knowledge of marine charts is necessary. In addition to the outline of land masses on charts, they also contain much other information in the form of symbols by which biologists can learn about the work areas. It is well to become acquainted with the most frequently used symbols, especially those signifying navigational dangers.

It is usually necessary to navigate a small boat and for this charts are required. For instance, if a sampling station is located in the centre of a bay, its exact location is needed, so whoever does the next sampling has the information enabling a return to that exact location. This determination is called position fixing. The position of drogues is also needed in current studies.

Positions on earth may be found in many ways, and the new electronic methods are very accurate but also costly and generally require large vessels or shore stations. A simple method uses a compass to determine directions from identifiable places on a chart. These are called bearings or position lines which may be transferred to a chart, and with at least two of these, a position may be determined.

Compass

There are many kinds of compasses such as gyro, astro and magnetic. For biological field work the magnetic type is most useful. Basically it is a magnetized iron needle, which, when allowed to swing freely will align itself with the north magnetic pole. The end directed towards the north magnetic pole is called the north seeking end. The circumference of the compass is divided into 360 equal parts called degrees and numbered from 0° to 360°, with the zero position designated as north. Some compasses may also be marked into 32 other divisions called compass points. The principal or cardinal points are north (0° and 360°), east (90°), south (180°) and west (270°). Between these are intercardinal points: northeast, southeast, southwest and northwest. Between these are further subdivisions which are now not often used except possibly for wind directions. The degree system for indicating direction is almost universal.

Although the north–seeking end of the magnetic compass points toward the north magnetic pole, it can also be deflected by iron or other magnets in close proximity. The difference in direction in which the compass is actually pointing and the direction of the north magnetic pole is called the compass deviation. In a fixed boat compass, the amount of deviation will vary according to the direction in which the boat is pointing so it is necessary to determine this for the various headings, for each compass and each vessel. In most inshore work with compasses, deviation will not be an important factor and in general can be neglected as long as it is remembered to keep the compass away from the engine or other metal objects.

The Terrestrial Sphere

The earth's axis is a straight line passing through the centre of the earth between the poles, which are the points where the axis meets the surface. The equator is a great circle on the earth's surface at an equal distance from the two poles at all points. The equator divides the earth into the northern and southern hemispheres. Both poles are 90° from the equator.

Meridians are great circles passing through both poles of the earth. As all meridians pass through both poles, they run north and south and are perpendicular to the equator. Of the infinite number of meridians that may be drawn on the surface of the earth, one, known as the prime or zero meridian, has been selected as the point of reference and is the one passing through the Royal Observatory at Greenwich, England.

The latitude of any place on the earth's surface is the angular distance north or south from the equator measured on the meridian passing through that place. Latitude south of the equator is termed south latitude and north of the equator is north latitude. Parallels of latitude are circles whose planes are parallel to each other.

Maps and Charts

A map or chart is a representation of a part of the earth's surface on a flat sheet of paper. Since the earth's form is globular any representation on a flat surface is an artificial construction and necessarily somewhat distorted. To satisfy needs of geometrical symmetry, equal distance and equal areas, certain methods of map construction are used. The two main ones are Polyconic and Mercator projections.

Mercator projection is mostly used for nautical charts since the meridians are made to be parallel straight lines equidistant from each other. The parallels of latitude are at right angles to the meridians and the course of a vessel or a bearing can be represented by a straight line.

When a map or chart is properly spread out, the top is north, the bottom is south, the right side is east and the left side is west. On Mercator projections, the graduated parallels of latitude on the right and left sides of the chart are marked off in degrees and minutes and fractions thereof and are, therefore, available for measuring distances (Fig. 57). However, distance between the parallels on a Mercator chart increase slightly in length from the Equator to the poles due to the distortion. In measuring distances, which may be done only with the latitudinal scales on the side of the chart, it is necessary to make the measurement directly opposite the area where the measurement is being made.

Latitude is reckoned in degrees, minutes and seconds from the equator to the poles. (One degree equals 60 minutes. One minute equals 60 seconds.) A nautical mile is equal to the mean length of a minute of latitude, which is 6080 ft (1854 m) as compared to a statute or land mile of 5280 ft (1610 m).

Longitude is reckoned from 0° to 180° east or 180° west of the prime or Greenwich meridian. The 180th meridian is known as the International Date Line which occurs in the Pacific Ocean. Contrary to parallels of latitude, which are equidistant from each other at all points, the distance between meridians or longitude varies with latitude, being greatest at the equator and narrowing to zero at the poles. Therefore, the longitude scale which is the horizontal one at the top and bottom of the chart, cannot be used for measuring distance

Longitude may also be measured in time. Each hour is equal to 15°, for the sun makes the 360° circuit in 24 hours.

In several places on nautical charts a compass diagram or 'rose' is printed (Fig. 58). There are two circles, and on the outer one the 0° - 180° line is parallel to the meridians and therefore points to the north pole or true north. The inner circle is usually slightly offset from the outer one in terms of its enumerations and the 0° - 180° line points to the north magnetic pole which is in the Hudson Bay region of Canada and to which all magnetic compasses point. The difference between true north and magnetic north is called the compass variation. If the north point of the compass is deflected east of true north the variation is said to be easterly and westerly if to the west. While the direction of true north is fixed (90° north latitude) the magnetic north pole tends to move its position very slowly. The compass variation on the particular chart is noted on the compass rose in addition to the annual change for which a correction must be made when a compass is involved, such as for steering. The variation also varies depending on the part of the world for which the chart is made (Fig. 58).

Bearings, Position Lines and Fixes

As stated above, simple navigational knowledge is necessary to determine positions, whether of a boat, a collecting station or current drogue. Position lines or bearings, which are simply compass directions, are lines joining known points on a chart. A common form of position line is when two known points identified on a chart are in line and these are said to be in transit. These could be a headland on shore and one end of an island. However, it is not known where on this position line the collecting station is located. This can be determined by means of another position line such as a compass bearing, and where the two lines cross will be its location. In large vessels, two sighting rods are mounted on the compass, allowing bearings to be taken, but in small boats, as used in field biology, small hand compasses are generally used. These also have sighting posts, often with a prism or mirror so that when the two posts are in line with the object of the bearing, the compass direction may be read.

There is an oyster study at Tanjung Pinang in the Riau Archipelago of

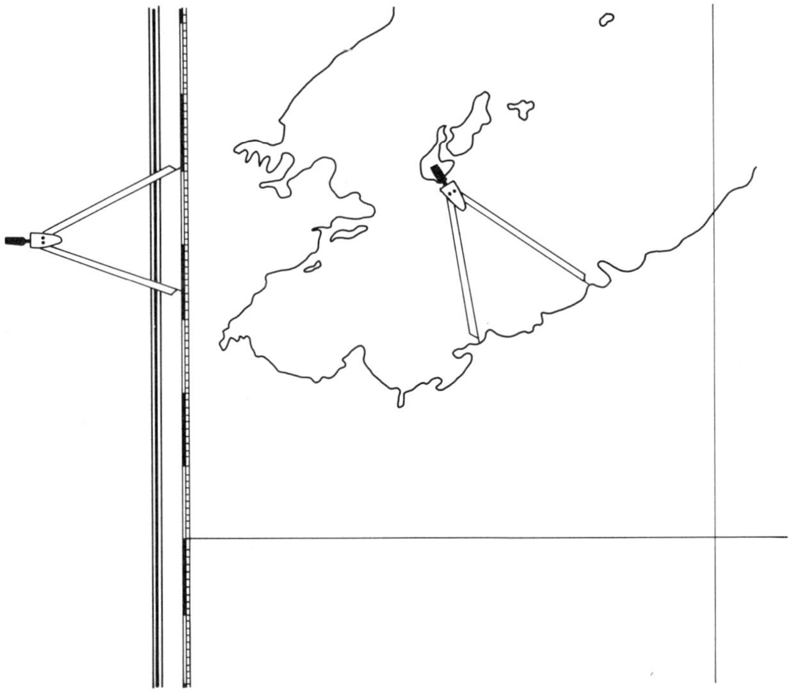

Fig. 57. Chart showing the use of dividers to measure distance by transferring the distance between two points to the latitude scale on the side of the chart (adapted from Canadian Hydrographic Survey chart 4354).

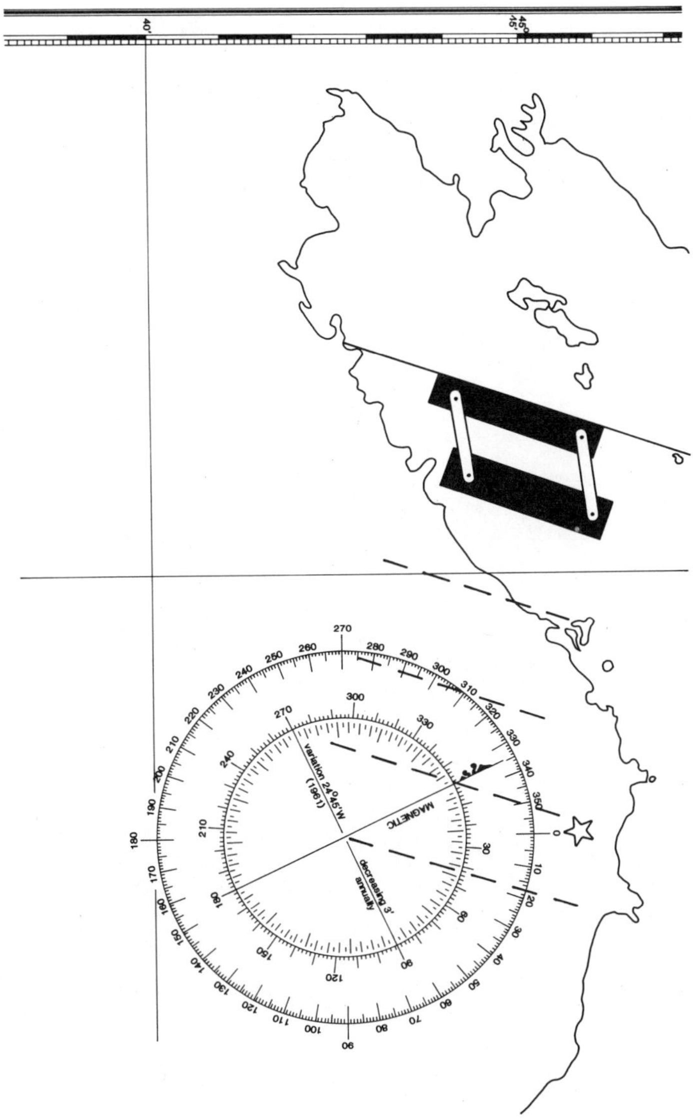

Fig. 58. Chart showing the compass rose and the use of parallel rule to transfer position line to compass rose to determine the bearing.(adapted from Canadian Hydrographic Survey chart 4354).

Indonesia and a plankton collecting station is being considered at the entrance of Bintan Bay (Fig. 59). In deciding the location it seems obvious to use a transit line from Los Island to Penyuengat. Another position line chosen was a compass bearing taken on the light house on Terkulai. This is determined to be 250° magnetic by transferring it to the compass rose by means of a parallel rule. On the compass rose the magnetic variation in this area is indicated as only 0° 30' E, which is so small it may be neglected. Thus, the magnetic bearing is practically identical to the true bearing. When the collecting station is visited to take samples, the two islands are placed in transit and the boat moved along that line until on the compass Terkulai light bears 250° magnetic. To plot the position on the chart the transit line is drawn and the reciprocal of the bearing of 250° is calculated (250° - 180° = 70°) and a line drawn 70° from Terkulai. The position is found to be 0° 58.6' North and 104° 23.8' East.

In the same way, it is possible to develop two or more position lines using compass bearings or two transit lines if this is possible. This is the case in Trincomalee Harbour in Sri Lanka. A collecting station in China Bay may be established by using Cod Point in transit with Mangrove Island and Middle Point with Mud Cove Pier. It is easier to relocate a position from transit line than from compass bearings. An additional transit line could be from Plantain can buoy and Powder Spit Light (off the chart). Three position lines give a more accurate location than two and provide a triangle called 'cocked hat' within which the actual position lies. The position here is 8° 33.9' North and 81° 72.2' East (Fig. 60).

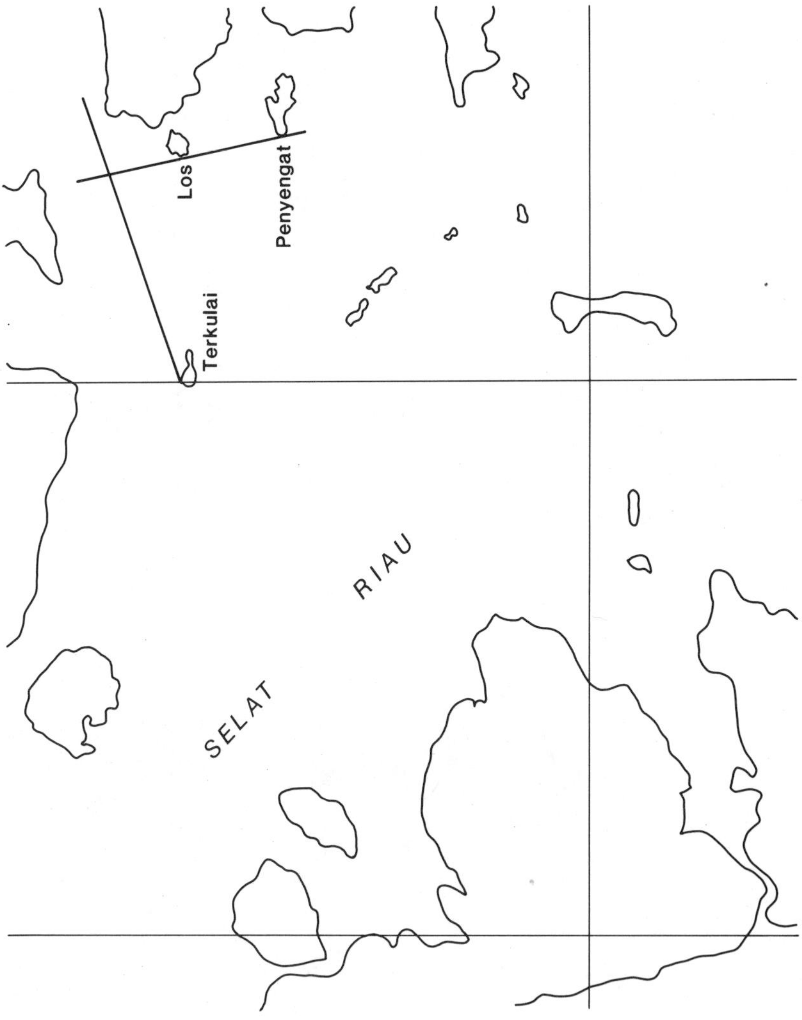

Fig. 59. Chart showing the entrance to Bintan Bay in the Riau Archipelago, Indonesia. Transit lines have been drawn to fix the position of a sampling station (adapted from British Admiralty Chart 3949).

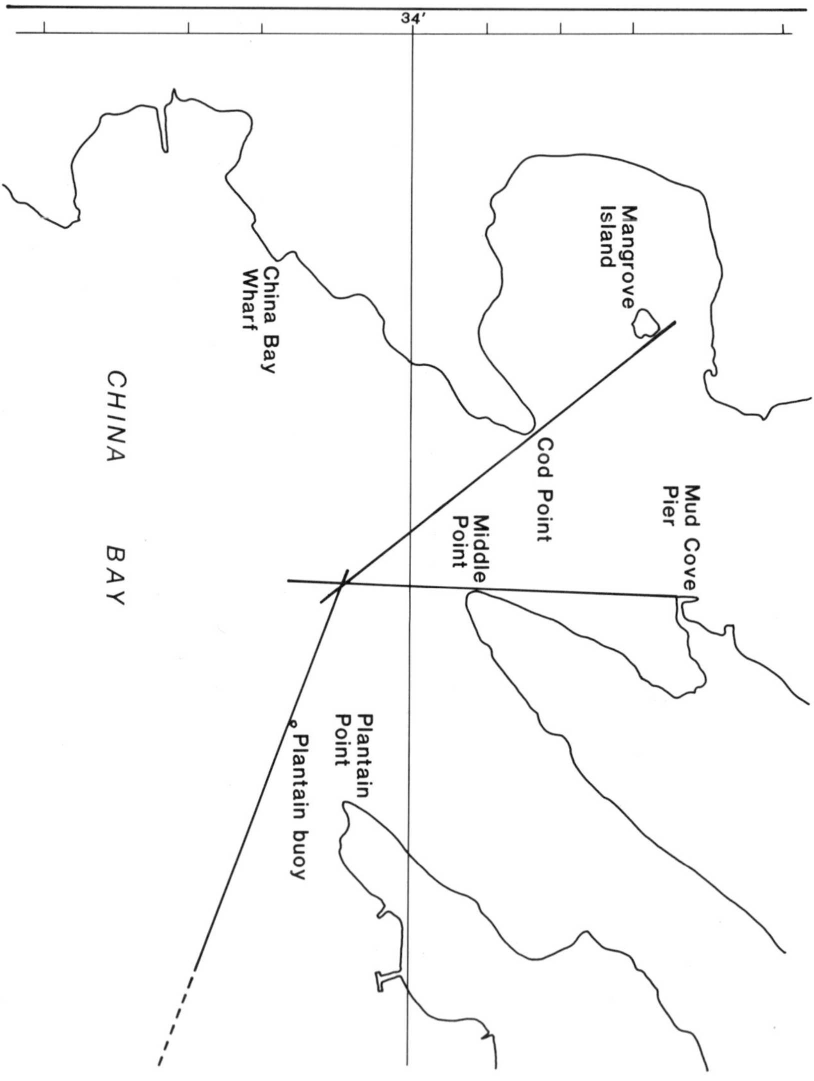

Fig. 60. Chart showing the Trincomalee Bay, Sri Lanka. Three transit lines have been drawn to fix the position of a sampling station (adapted from British Admiralty Chart 3949).

CHAPTER 6

OYSTER CULTURE

As discussed previously, oyster culture is probably the oldest form of bivalve culture. Until recently, it has been the most widespread and largest form of bivalve culture. The reason is that oysters are more suitable to artisanal culture or a form of semi-culture involving collection of wild seed. There are actually few basic ways in which to culture oysters but an infinite number of variations on these main themes is possible. The following is a list of the basic types of oyster culture.

1. Bottom
   (a) intertidal
   (b) subtidal

2. Rack
   (a) tray
   (b) string
   (c) stick

3. Raft or long line
   (a) tray
   (b) string
   (c) stick

4. Stake

In a situation where a development project is set up to evaluate culture methods, it may be necessary to start equiping a laboratory and workshop. To assist in this, a suggested list of equipment and supplies is provided in Appendix I.

BOTTOM CULTURE

Bottom culture is the natural extension of the way in which unexploited oyster populations are found on natural beds. In nature, populations were maintained by spatfalls on parent oysters or dead shells. This often creates oyster "reefs" with generations layered one above the other, with those oysters on the lower layers finally dying either by smothering or competition for space and food. A similar, though not so extensive, situation exists on mangrove roots but, as soon as the layers become heavy enough, they fall off the root and the process is repeated.

The object of the culture process is to replace the reef, with its mixture of age classes, with a situation where age classes are kept separated on different grounds. Some oyster growing areas are located in breeding areas, so there is the possibility of a multiplicity of age classes which can cause difficulties in harvesting. Often the best seed producing areas are partially separated from growing areas by hydrographic factors. Such a situation occurs in Virginia on the Atlantic coast of the USA, where the seed collected in the James River is used on growing grounds not far away.

The ideal situation is where breeding grounds are completely separated from growing areas. This was so with the Japanese oyster (*Crassostrea gigas*) introduction to the west coast of the USA and Canada. Initially, the Japanese oyster did not breed in the new environment, and there was an annual importation of seed. Thus, it was possible to keep the year classes completely separated. This made productivity measurements simple, for the exact number of seed planted was known and could be compared to the number of oysters harvested without the problem of the contribution of natural spatfalls.

The ideal situation seldom occurs, but there is always a possibility of growing oysters in areas where they do not breed naturally. It should be mentioned that eventually the Japanese oyster did breed in a few selected areas in Western North America, but these do not generally interfere with culture in the non-breeding areas, they do furnish a supply of seed that is no longer required from Japan.

Intertidal Bottom Culture

In addition to those elements of site selection already discussed, there are three main characteristics of beaches useful for intertidal oyster culture:

1. Bottom consistency
2. Tidal height
3. Protection from wave action

Bottom Consistency

The bottom must be firm enough to support the oyster in position after planting. The bottom may be of rock, sand, gravel, mud or any combination of these. Level sandstone type rock is suitable but must be relatively even to permit easy harvesting. It has the disadvantage that slight wave action can move the oysters, either rocking to knock off shell growth or piling them into windrows. Boulders may be used to construct low fences as wave barriers on stone ground.

Pea gravel (1 cm in diameter) forms the ideal bottom surface. It is generally level, gives adequate support, permits development of well–shaped oysters and harvesting presents no difficulties. There may be some tendency toward movement by wave action but this may be reduced by planting the oysters quite thickly so they support each other.

Sand is generally firm enough to support oysters but, particularly in backwaters, there may be areas that are too soft because water content of the sand is too high. Burrowing shrimps may also soften a sandy bottom by increasing the water content with their tunnels to the point where it cannot be used. Riffled sand is firm enough, but riffling indicates a considerable amount of wave action which will cause shifting sand and bury the oysters. Look for stability in a sand bottom.

There are probably more oysters grown on mud than any other substrate because of its ubiquity. The difficulty is assessing the limit of softness of mud that will support oysters. A rough rule of thumb requires a 60– to 70– kg person to sink not more than 10 to 15 cm when walking over the bottom. There is a tendency for oysters grown on mud to develop an elongated shape with a smooth shell surface.

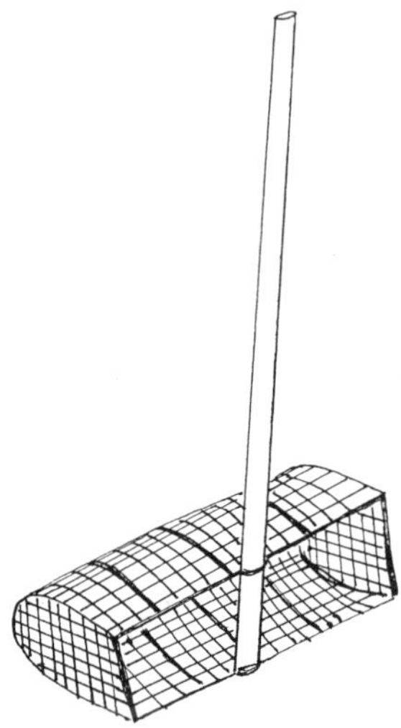

Fig. 61. Basket rake, similar to those used in blood cockle culture in Malaysia, that can be used to harvest oyster or mussels from shallow water.

Tidal Height

A reasonable amount of intertidal bottom between 2 and 4 ha should be uncovered to provide enough area to make culture worthwhile. In situations where the tidal range is small, there may be additional ground marginally subtidal that may be used. If this is no greater than about 1 m deep at low tide, it will be possible to culture by wading, using a basket rake (Fig. 61) for harvesting.

With an extensive tidal range, 3 m or more, assessment should be made of the tidal levels at which predators such as crabs, drills and seastars occur, as well as zones in which barnacles and other fouling organisms settle. It may be assumed there will be a gradient in growth rates down the beach, with the greatest rate at zero tidal level. The question is what upper tidal level will provide a satisfactory growth rate while considering the fouling zones and how serious predator problems may be.

In one system of bottom culture, it is the practice to plant seed high in the intertidal zone, because it is there that predation and silting mortality are low, thus compensating for reduced growth rate at that level. In the second year, the seed is relayed to a low tide area to provide rapid growth, and in the third and final year moved to a medium tide level where good fattening often occurs. This is an example of the manipulation possible, making the best use of the tidal range. The situation in the tropics is different because of the reduced time frame, approximately 1 year instead of 3 or more to a market size, and the problem of heat stress with long periods of air exposure.

Protection From Wave Action

Wave action, if of sufficient magnitude, can cause movement of the oysters along the bottom, and in the worst case, they will be found in rows along the high tide mark. Also, wave action can cause oysters to be buried by sand or mud or make harvesting difficult. Tumbling knocks off fragile shoots of new shell growth. Turbidity from wave action causes oysters to expend additional energy separating food from the sand and mud particles in the feeding process. In some cases, the turbidity may be enough to cause cessation of feeding.

Only experience can determine the amount of permissible wave action, for it depends partly on the slope of the beach and whether the waves are of the long ocean swell type or short wind induced waves. Duration of the wave action is also a factor whether they be steady as with normal trade winds or only occasional occurrences. Some indication may be obtained from the distribution of flotsam on the beach. Unless it is on a surf swept beach it usually shows a riffled surface from wave action. Potentially valuable intertidal bottom may be protected from wave action by floating breakwaters, but this is a last resort.

Predation

Bottom cultured oysters are quite susceptible to predation, since most oyster predators live on the bottom. Among these are gastropods (drills), seastars and fish. The former two have been dealt with in another section.

The main fish predators are rays which, with their powerful jaws, are able to crush oysters. It may be possible, according to the abundance, to reduce the population in the immediate area of an oyster bed to acceptable limits by fishing. In some oyster areas beds are fenced to keep rays out. Fences up to 2 m high are usually required. An alternative is to stud the bed with sharp stakes about 30 – 50 cm long. If placed close enough together the flat rays, whose mouth is on the under surface, are unable to settle down on the oysters. The fence system has been used on the west coast of the United States in Humboldt Bay and the stake method in France, particularly in the Arcachon area.

Fouling

One advantage of bottom culture, particularly of the intertidal method, is that fouling is generally minimal. It is here that fouling organisms are usually distinctly zoned, permitting placement of the plantings to escape the major effects.

The barnacle is one of the main intertidal fouling organisms but is often zoned quite high in the intertidal area. Barnacle settlements do not seriously affect growth of the oysters and are mainly a nuisance. Encrusting or finger-like sponges may attach to oysters, particularly in the lower tidal reaches especially where the ground is moist or where there are shallow pools. Other major intertidal fouling organisms are algae. There is virtually no way to avoid algae. If the operation is of artisanal size, it may be possible to rake the algae, or most of it from the oysters.

There usually are a number of other fouling organisms, but they are generally small and do not occur in sufficient abundance to form a hazard. More detail on the groups of fouling organisms is given in the general section on fouling.

Seeding

Once an area has been selected for intertidal bottom culture, it is necessary to allocate areas for seeding and areas for growing. Two main sources of mortality (excluding predation) for oysters on a piece of seeded cultch placed on the bottom are competition for space and silting. Little can be done about competition. If there is a heavy set, the oysters will thin out through competition and there will usually not be a serious problem. However, ground selection can eliminate or reduce silting mortality. The degree of silting will be indicated by the softness of the ground, so it is obvious to select as firm a bottom as possible. This is most often found at higher tidal levels; thus, it may be necessary to balance possible reduction in growth rate associated with higher tidal levels with the reduction in seed mortality. Ideal bottom is one consisting mainly of pea gravel with adequate protection from wave action. It is possible to plant seed in one area and transfer to another for final growth.

The following factors should be considered in the selection of seed ground.

1. Amount of suitable ground at various tidal levels in proportion to the amount of seed to be planted.
2. Bottom consistency. Firm bottom reduces mortality and produces cupped oysters.
3. Seed at lower levels is more susceptible to attack by predators such as seastars, crabs and drills.
4. Time available to work on seed at low tide levels is less than at high levels.

Tidal levels	% Mortality	
(m)	Tray	Bottom
1.5	33.8	38.9
1.0	43.8	64.3
0.5	29.3	70.6
0.3	35.6	77.8
Average over all levels	35.6	62.9

Table 5. Comparative mortality of *Crassostrea gigas* seed on trays and on bottom (modified from Quayle, 1988).

An example of the effect of silting mortality is demonstrated in an experiment with *Crassostrea gigas*. Sixteen random groups of 100 pieces of spatted shell cultch were alloted randomly to 8 trays (1 m^2) held off the bottom and 8 plots (1 m^2) on the bottom. These were arranged in 4 groups of 2 trays and 2 bottom plots at each of 4 tidal levels between the 0.3 m and 2 m tidal level, for a 7–month period during the summer growing season. The bottom was muddy but with a gradient in softness between the two tidal levels. Mortalities of these groups are shown in Table 5.

The survival gradient from the upper level to the lower is quite evident. The factors may be described as follows:

Bottom mortality = natural mortality plus competition for space plus silting = 62.9%

Tray mortality = natural mortality and competition for space = 35.6%

Silting = the difference = 27.3%

Substituting the experimental values and cancelling out the common

Spat count (100 shells)	Number surviving	% Survival	% Mortality
2386	1236	51.4	48.6
2269	1186	52.3	47.7
2133	1224	57.4	42.6
1956	1325	67.7	32.3
1953	1194	61.4	38.6
1489	1087	73.8	26.2
1330	952	71.5	28.5
1238	994	80.2	19.8

Table 6. Spat counts and survival per 100 shells held on trays (from Quayle, 1988).

factors of natural mortality and competition, an estimation of 27.3% is shown for silting mortality. The experimenters saw no evidence to suggest appreciable natural mortality, so tray mortality of 35.6% may be accepted as due to competition for space. In Table 6 is shown spat counts per 100 pieces of shell cultch on each tray in the same experiment where there was no silting mortality. The data indicate that competition for space is least with the lowest spat counts per unit of area. Notice how the number of spat tends to even out. The starting range of about 12 to 24 per shell ends up with 9.5 to 13 per shell.

Spreading

Seed may be spread either from a boat, or directly on the ground at low tide. In either case, the ground must be carefully marked by stakes driven into the bottom. Often, boat seeding requires some redistribution afterwards.

There are no strict rules on the density at which seed is to be planted as it depends on many factors. The amount of seed ground available relative to total area may influence the density, as will the level of wave action. Dense planting will, to some degree, prevent movement of seed by wave action. If the seed is to be transplanted, it is more economical to do so from a dense planting, as pick up costs are less than from a sparse planting where raking may be required. A dense planting may be forked directly into a container, whether it be a wheelbarrow, a basket, a boat or top float.

In temperate zones, using shell cultch with about 15 spat/shell, a density of 50,000 shells/ha is an average planting. Reduced to smaller units this is equivalent to 4 or 5 pieces/m^2. With maximum survival, this will result in 60 to 70 mature oysters/m^2 but in actual practice on average ground where silting mortality is not too high, there is survival of about 20 mature oysters. On an area basis, this is equivalent to 200,000 oysters/ha. The tonnage of oysters in the shell and meat yields will depend on size. If the size of the oysters is such that there are 30 whole oysters/litre the yield will be about 60 t/ha or 6300 litres of meat.

This planting density is not particularly heavy and plantings four times as high may be made depending on the circumstances, relative to availability of ground, its type and culture sequence. Moderately dense planting does not affect growth significantly.

Alternatives in the utilization of seed ground are:

1. Plant at a low density such that when clusters are separated on the spot, the seed will represent the number of oysters that may be matured in the area without transplanting.

2. Plant at a high density and transplant all the oysters to other growing ground.

3. Plant at a high density and transplant only some oysters, leaving the remainder to mature on the spot.

As with all beginning oyster culture efforts, seeding operations are initially one of experimentation with the variables of bottom type, tidal height, density and labor costs.

Cluster Separation

The seed on most cultches will grow into clusters. A shell cultch 8 to 10 cm in length with a spat count of 25 to 100 or more will usually yield about 20 oysters on moderately firm bottom. The seed are attached to each other and to the mother shell, and as they grow there is a tendency to elongate. This process will become progressively worse unless the cluster is separated, preferably into single oysters, but doubles or triples are satisfactory. In temperate waters, *Crassostrea* may attain a length of 4 to 5 cm after one summer of growth in a cluster. This is an appropriate size for separation, for they will have grown away from the mother shell, leaving minimal attachment at the umbone. In the tropics, separation should take place at about the same size, whenever that occurs.

Breaking may be accomplished by pulling clusters apart by hand (using gloves) or by tapping the mother shell with a flat iron bar or prying the oysters apart with it. Another useful instrument is a rake (Fig. 62) with an iron knob welded to the bar. The rake is used to turn the cluster over to expose the mother shell which is then tapped by the metal knob, used as a hammer.

Declustering operations may result in a mortality of up to 25%, depending on the state of the shell in the cluster and how carefully the breaking is done. Mortality is in part compensated for by better shape and more rapid growth in the survivors as well as speeding up the shucking process. It is difficult to shuck oysters in a cluster.

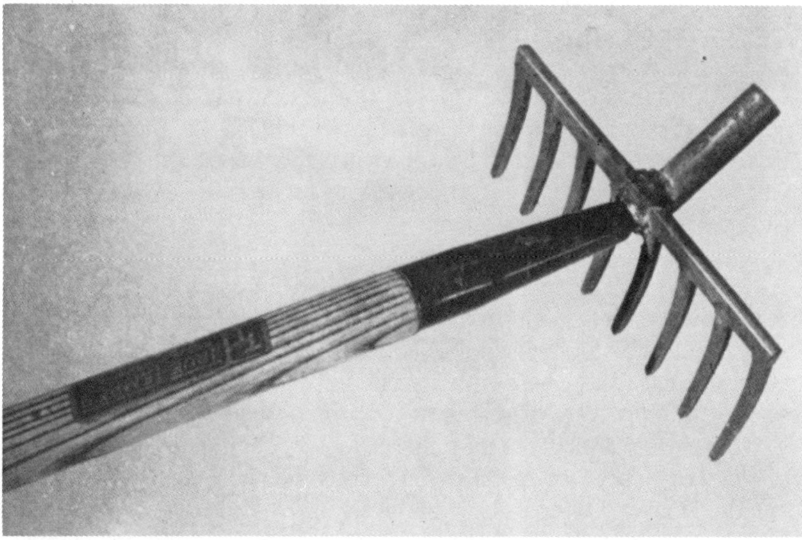

Fig. 62. Cluster–breaking rake (from Quayle, 1988).

Excessive mortality may be caused by:

1. Attempting to obtain too many single oysters.
2. Carelessness
3. Seed too small or too large

If the oysters are too small, they are easily broken. If too large, they will be firmly attached to each other and excessive shell damage may result. Experience is the best guide to timing the declustering process. The cost of separating clusters is variable, but 1000 clusters should require no more than 2 to 3 man-hours with hand breaking.

To reduce the size of clusters, the cultch may be broken or cut into small pieces before it is planted. This is done when the seed is no more than 2 mm in diameter. Theoretically, competition for space on the cultch is reduced, for the perimeter is increased, allowing the seed to grow out from the cultch edge. However, this gain may be lost by the tendency toward increased silting mortality, with small cultch pieces. Large scale studies have indicated little difference between small and large pieces of cultch when grown on similar ground.

Growing and Harvesting

If seed has been planted at a low density for growout without transplanting, clusters should be broken when the oysters reach 3 to 5 cm in length. Tests on a few clusters from time to time will determine the approximate time of readiness for this operation. As clusters are broken, individual oysters are distributed evenly over the bottom.

There is virtually no information on bottom culture in the tropics partly because most culture attempts have been done off bottom. It is doubtful if there are extensive areas of suitable bottom because most potential growing areas are in estuaries where bottoms tend to be muddy.

After seed planting and then cluster separation, little effort is required until harvest, depending on the seeding plan which may require some transplanting. However, careful observations should be made on growth rates and survival in various parts of the growing ground. Fouling and possible predation should be monitored. Towards the end of the growing period, checks should be made on the condition factor. In the initial stages of the operation the shell size to meat ratio should be calculated at various times as the market period approaches to determine the optimum size for marketing. With the short growing season in the tropics, it will be necessary to plan for continuity of supply and this will relate to seed availability, market season if there is one and availability of ground.

This discussion is confined to the problems of small scale culture where all operations are conducted manually. Mechanization is probably economical only on large scale–oyster farms.

If seed has been planted at a high density with a view to transplanting to growing ground, the appropriate method is to move it before breaking the clusters. Hand picking, raking or forking the oysters into the transfer receptacle (buckets, baskets, wheelbarrow or boat) will automatically break some clusters. The others may be broken on the new growing ground.

The raking process, in which an ordinary garden rake suffices, moves the oysters into piles or windrows so they may be picked up with a fork — usually a pitch-fork type with tines spaced to hold the oysters as they are lifted.

In some instances, it may be expedient to move the oysters a second time to fattening ground if, after trials, certain areas are recognized as being exceptionally productive. Another reason for a second move may be to bring the oysters close to the base of operations, to a processing plant or for depuration before marketing. With a relatively brief growth time for tropical oysters, more than one move may be impractical.

Since in intertidal bottom culture the oysters are not always accessible during neap tidal periods, a system of storage is required to keep the market supplied. Storage ponds such as those used in the Netherlands are costly. In some cases oysters can be stored for brief periods at a relatively high tidal level. An alternative used in some countries is the sink float. This is simply a float with the floor submerged below the surface of the water. The oysters may be loaded to form a layer of up to 30 cm or so. A small area of float can accommodate a considerable number of oysters. This system of storage also allows the oyster to be cleansed of sand and mud particles. The main problem is to find a sink float location where the bacteriological water quality is satisfactory. One advantage of sink float storage is that the float may be

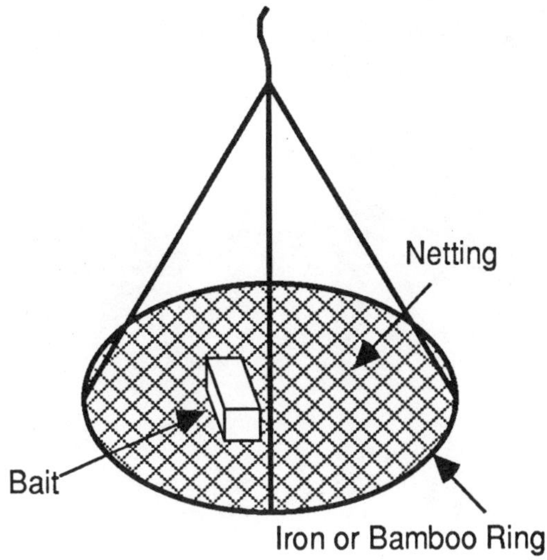

Fig. 63. Baited ring net for starfish control.

moved, for instance from the harvesting area to the unloading area, again providing quality of the water through which it passes is satisfactory.

Subtidal Bottom Culture

As with intertidal bottom culture, one of the main requirements is a firm, stable bottom (with minimal siltation). Instead of tidal height, water depth must be considered. Deep water makes harvesting difficult, so a relatively shallow water depth is required, probably not more than 3 to 5 m. This permits brief examination of plantings by free diving. Wave action will be an important factor only in very shallow areas (1 m deep). Instead, the speed of the tidal current is of concern, for a rapid current can move oysters along the bottom as readily as wave action.

To bottom consistency, water depth and current speed, must be added predators and fouling as factors in considering subtidal culture. Predators, especially seastars and drills, are not readily controlled as in the intertidal situation. Drills may be trapped and, with some difficulty, seastars also. For

Fig. 64. Typical use of tongs for oyster harvesting.

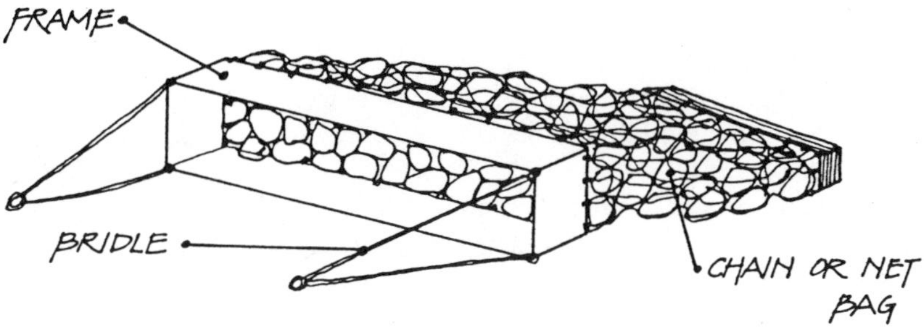

Fig. 65. Oyster dredge (from Quayle, 1980).

the latter baited ring nets (Fig. 63) work well, but they must be tended fairly regularly for when the bait disappears the seastars will leave the net. In Long Island Sound in Eastern United States, where subtidal oyster culture is normal, seastars are controlled by vessels towing a series of string mops along the bottom. The sea stars become entangled in the mops which are lifted and dropped into vats of boiling water. This is a costly but necessary operation. An alternative is to spread a thin layer of quicklime over the bottom. This destroys seastars but does not harm the oysters.

Fouling in the subtidal is always more extensive than in the intertidal zone, although generally less than in suspended culture. Subtidal fouling is difficult to control and is one possible negative factor of this method, especially in the tropics.

Otherwise, the culture system is similar to intertidal culture with similar planting densities. The main difference is in harvesting. Dip nets, tongs (Fig. 64) or dredges (Fig. 65) are used. Subtidal culture is not a typical artisanal system and, at the present time in the tropics must be considered as a last resort after other systems have been investigated.

RACK CULTURE

Rack culture is generally carried out in the intertidal area and is best done where there is at least a modest tidal range. Racks may also be placed partially subtidal at a level that allows work to be done by wading. Advantages of rack culture are:

1. Independent of bottom consistency
2. Growth rate more rapid than on the bottom
3. Fewer predator problems
4. No silting mortality.

From racks, any number of culture devices may be hung. Among these are strings, trays, lantern or pearl nets, tubing or sticks. Racks may be erected either on firm or very soft bottom. The latter bottom type is usually more extensive and available than the former. Since oysters on racks are off the bottom, there is no danger from silting mortality unless turbidity is excessive. Oysters off the bottom grow more rapidly than those cultured on the bottom, presumably because food is more available and less energy is expended in separating out food from silt which is always more prevalent along the bottom. The danger from predators is not completely removed but control is easier. Drills are the main culprits but they must scale the posts of the racks before they can reach the oyster. On muddy ground a problem that arises in rack culture is the buildup of silt beneath the racks. Depending on the amount of silt and the current configuration, this build up may be quite significant, and the racks should be moved before the silting becomes excessive.

Rack Construction

There are 4 main types of rack construction:

1. Single beam
2. Tripod
3. Cross beam
4. Parallel beam.

Single beam. This consists of a single beam placed and secured to the top of a series of posts driven into the bottom (Fig. 66). Material for rack construction generally available in the tropics is bamboo, mangrove or other local woods such as bilian. The diameter of the posts and beams should be in the 8 to 10 cm range. Posts vary in height depending on the type of culture material to be used. For trays, sticks and horizontal strings, posts need be no

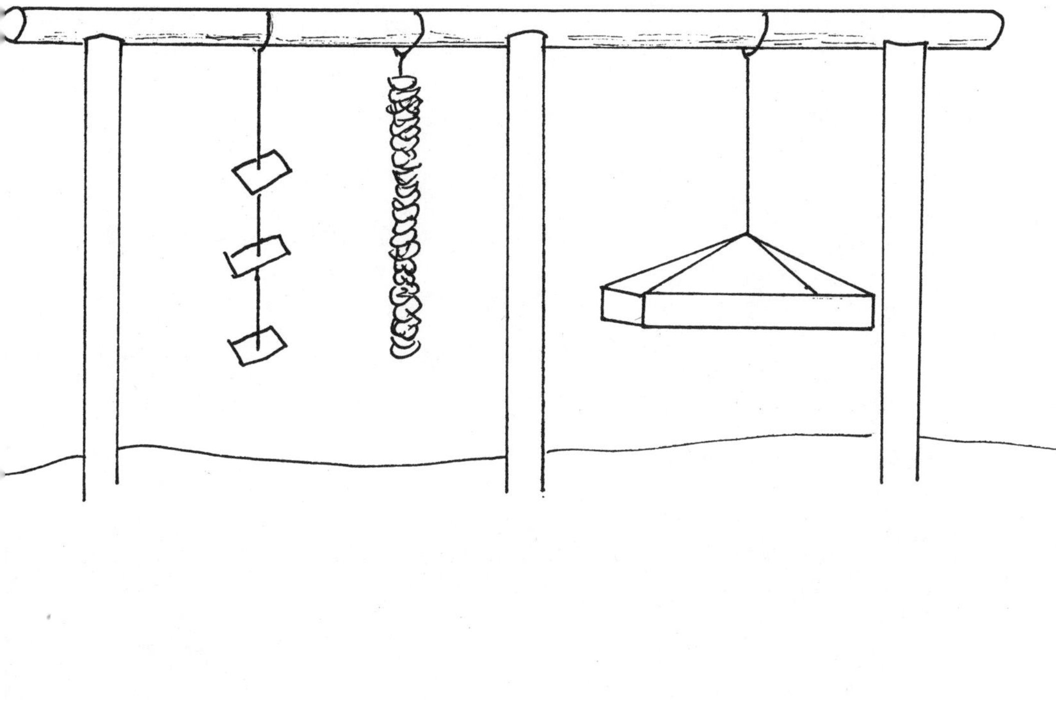

Fig. 66. A single–beam rack that can be used to hang monitoring plates, cultch or trays.

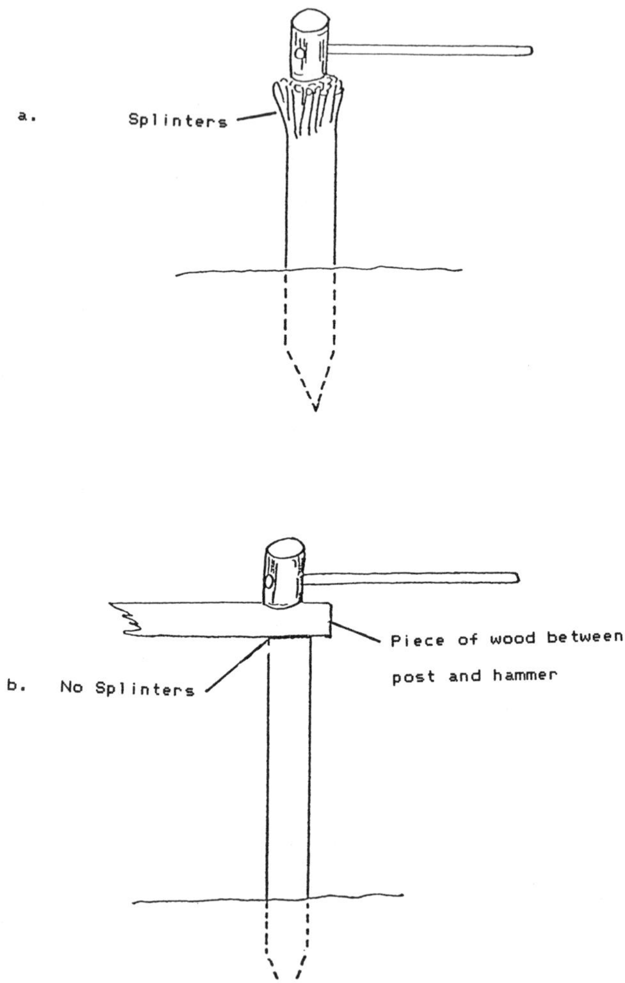

Fig. 67. a) The wrong way to drive a post results in splintering. b) The right way leaves a solid post for rack construction.

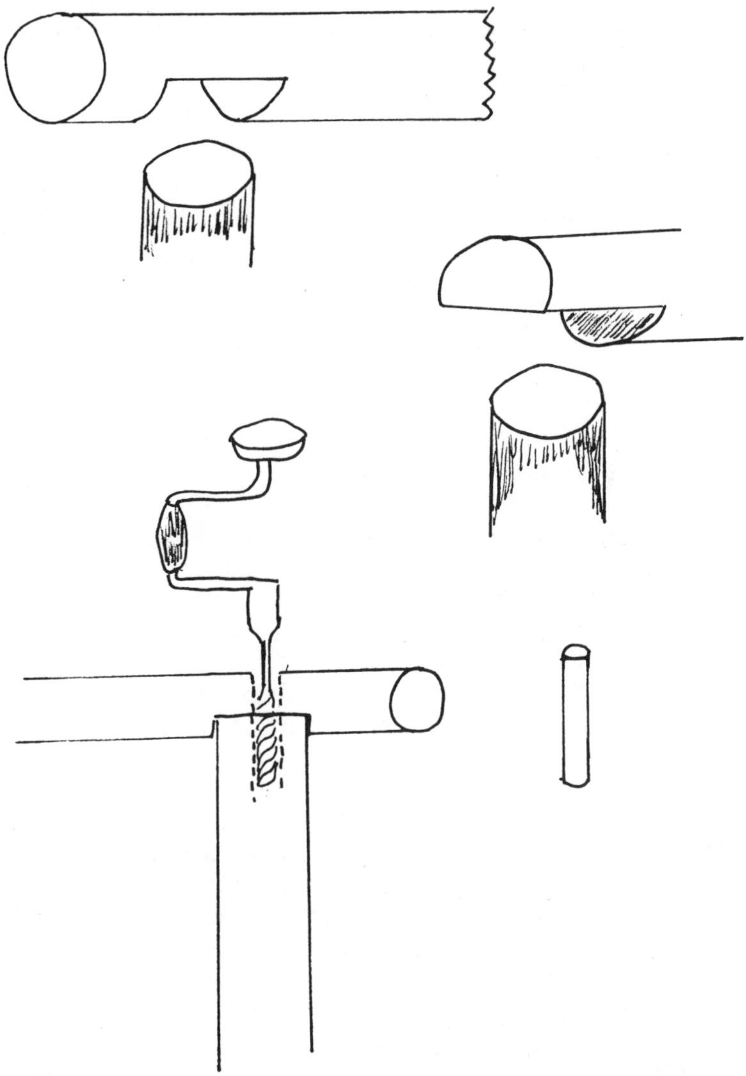

Fig. 68. Beams can be notched or lapped. The beam can be secured to the post with rope or wire or with a peg as shown in the lower disgram.

more than 60 to 70 cm above the bottom, and driven into the ground nearly as far. For string culture, the aboveground part of the posts should be longer than the length of the strings. The bottom end should be pointed and a piece of flat wood placed on the other to take the force of the hammer blows to prevent splintering (Fig. 67).

The length of the beams will also depend on the type of oyster culture material, but 3 or 4 m will be satisfactory for most situations. It must be remembered that as the oysters grow, weight increases and it is a common error to use materials only strong enough to hold the initial weights.

For bamboo posts and beams, nails are difficult to use, so joints must be tied with flexible lianas, wire or cord which should be not less than 3 to 5 mm in diameter. With mangrove, construction posts and beams may be notched (Fig. 68) and either pegged, nailed or tied.

Tripod. Another form of beam support, useful if strong enough posts are not available or if the bottom is too hard to drive posts, is the tripod system. For the best support three poles are spread an equal distance apart on the ground and brought together near the top and tied (Fig. 69) at the appropriate height. One tripod is needed at each end of the rack but the intermediate supports need only be with two poles. This rack is usually used for string culture where a greater distance from the bottom is needed.

Cross Beam. The cross beam rack is a slightly more complicated rack system because of the number of connections. On top of single posts a cross bar is placed and supported at each end by angular braces attached to the post (Fig. 69). The long beams are placed on the end of cross beams. This is the system used in Cuba where the culture consists of suspending brush tips of mangrove trees with monofilament nylon from the beams. Oyster seed is collected on the brush tips and allowed to grow to maturity *in situ*. This type of rack can not support a lot of weight.

Parallel Beam. The commonly used parallel–beam method is simply doubling of the single beam system with two lines of posts and beams connected by cross beams (Fig. 70). Construction methods are similar to that for the single beam. This type of rack is particularly good for trays.

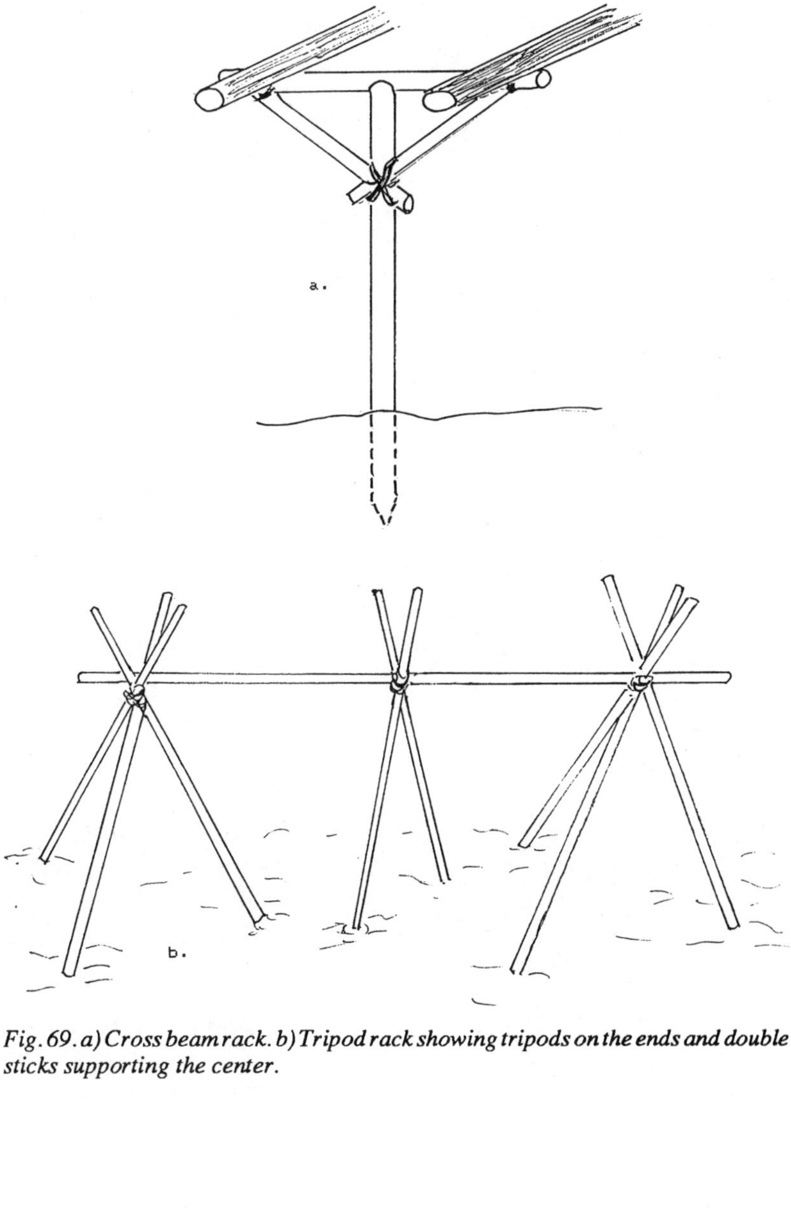

Fig. 69. a) Cross beam rack. b) Tripod rack showing tripods on the ends and double sticks supporting the center.

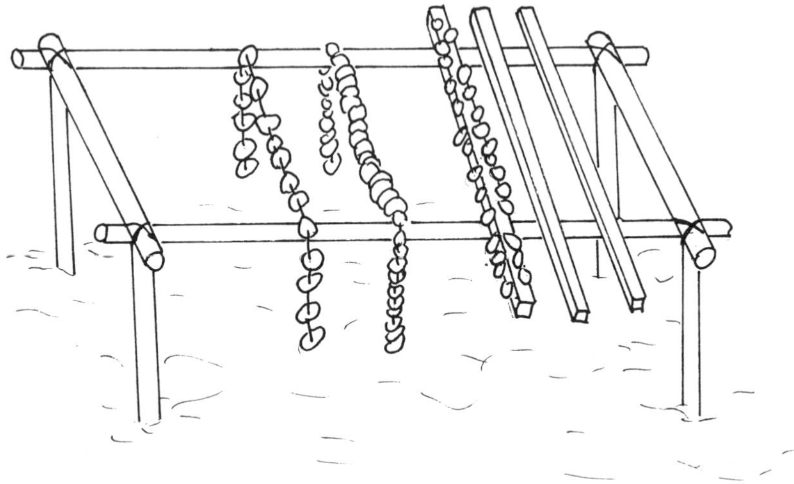

Fig. 70. Parallel beam rack which can be used for strings, sticks or even trays.

Tray Culture

Trays have been used for oyster culture for a long time, although not extensively probably because of the costs and the vast numbers required for a large production. They have been used in Norway, where there is no available bottom and where the European flat oyster (*Ostrea edulis*) does not lend itself well to other types of culture. Trays have also been used in Japan for pearl culture. They were normally made of iron or wooden frames covered with mesh, but in recent years plastic ones have been designed and made specifically for oyster culture. Trays have been used extensively by oyster biologists, particularly for comparative studies since they provide a uniform platform and can be placed in identical situations as at specific tidal level or depth at various sites.

The high cost of producing tray oysters requires a high price for the product which, in turn, requires high quality. High quality is too often equated with uniform oval shape for the halfshell market but with little reference to the quality of meat (within limits). However, off–bottom oysters, whatever the shape, produce meats of higher quality flavour and condition factor. The simple culture technique requires single spat to make most efficient use of the surface area. This is difficult with clusters.

Seed for trays can vary in size from very small (2 to 10 mm) single spat from hatcheries to almost any size, though most often not greater than 3 or 4 cm. With small hatchery spat, it is necessary to place a small mesh liner in the trays, for growing trays have fairly large meshes or apertures to allow adequate water circulation to the centre of the tray.

Very small spat have a tendency to concentrate in corners of the tray owing to wave action so they must be respread quite frequently. If floating trays are used for hatchery spat, the wave action will usually move the seed back and forth regularly. There is a tendency in early stages of growth for some species to grow together and attach to the tray, so this requires attention. These characteristics have largely precluded the use of single spat from hatcheries, and other sources of larger spat are required.

To make optimum use of tray space, the oysters must be quite tightly packed into the tray, preferably with the hinge down. Tight packing may reduce growth rate somewhat and it is a compromise between that and space utilization. An extra month or so of growth time to market maturity is usually less costly than extra trays. Tight packing also reduces movement in the tray.

As with bottom culture, the seed may be planted densely at first in the trays and then transferred to other trays as they grow and fill the first tray. The seed may be planted less densely and allowed to mature in the same tray.

Trays may be suspended by a bridle from a single beam rack or placed on a double beam rack. Some saving is obtained by attaching it to a single, short cross piece the length of the tray to support it on the double beam rack. In this case it may be necessary to tie the tray down.

As a result of the protection afforded by the tray, predation is generally not a significant factor. If necessary, a lid may be attached particularly to keep out crabs, but complete enclosure also provides surface for fouling which, except for barnacles, is not too great in the intertidal zone. Placing several herbivorous snails or small crabs inside the tray also assists in biological control of fouling.

The most difficult management problem for trays, especially subtidal trays, is to balance the need for a mesh size sufficient to hold the oysters against the need to have good water flow. Small mesh openings restrict water flow particularly when they become fouled. Thus, the oysters should be moved to larger and larger mesh size as they grow. This often requires grading of the oysters, since there will be considerable variation in size. If fouling can be controlled the need for mesh size larger than 1.5 to 2 cm is reduced.

String Culture

String construction for seed collection with shell, bamboo, etc. has already been dealt with. Strings for growing are constructed in the same way, except the individual pieces of cultch are separated by a space of 20 to 30 cm, depending on the size of the oysters grown. The method of attaching cultch to strings depends on the type of string and the type of cultch. The spacing method depends partly on the string material and on whether or not the cultch piece has a hole. If it has a hole, it may be strung on wire, twine, rope, etc. with a tube spacer about 15 mm in diameter of bamboo or plastic water hose (Fig. 71). If the stringing material is wire (No. 12 or 15 galvanized), a twist in it (Fig. 72) will keep the cultch pieces apart. Knots tied in rope can be used but this is a time consuming method.

These are not the only methods. New methods or adaptations should always be sought. The insertion of separators such as bamboo or plastic pipe is time consuming and costly in comparison to other methods where all of the cultch pieces may be threaded on the string at one time. In the latter case, the separation is made as either the group or single pieces of cultch are slid along the string. This is the method used on wires where a twist in the wire will hold the cultch in place.

If the cultch pieces are not punched with a hole, separation is accomplished by opening the strands of a two strand rope and inserting the cultch (Fig. 72). This was the original method used in Japan with tarred straw rope but it eventually gave way to wire strings with bamboo spacers. However, two strand synthetic 3–mm diameter rope is now available. Insertion between the strands of rope is rapid but it means the spat must be very small and the whole growth period must take place on the string, as the cultch with large oysters intact can not be easily removed. This is not always advantageous.

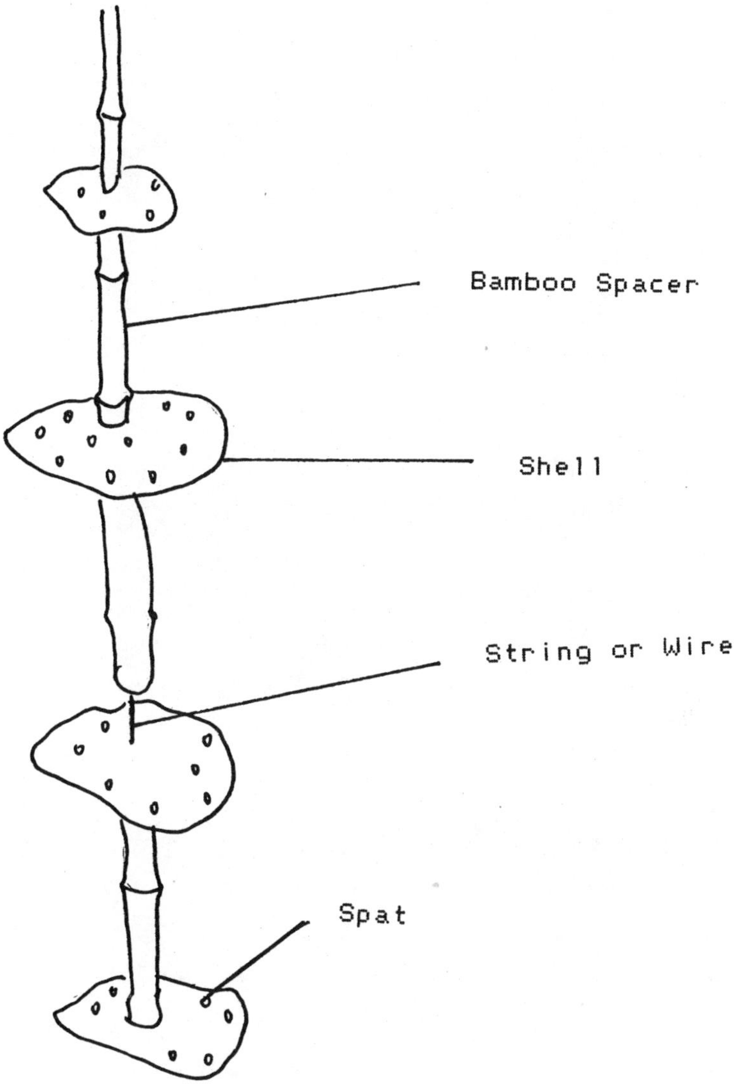

Fig. 71. A string ready for grow out with spatted cultch separated by bamboo sticks.

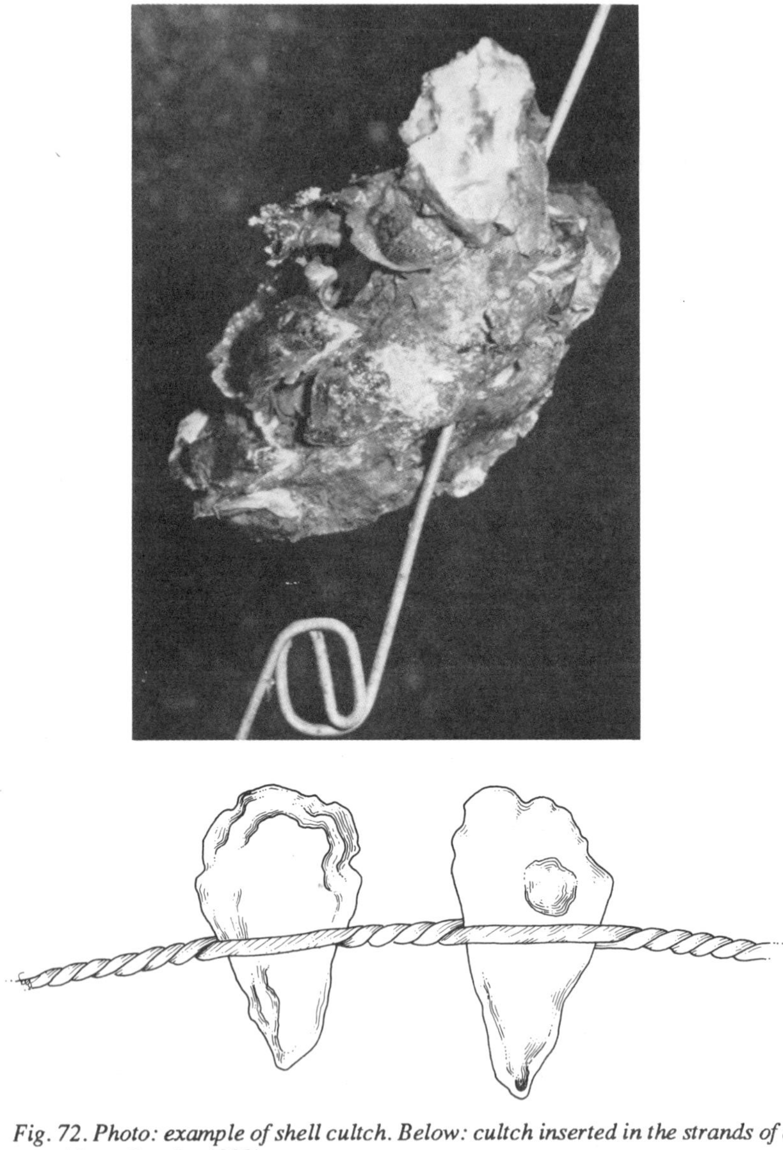

Fig. 72. Photo: example of shell cultch. Below: cultch inserted in the strands of a rope (from Quayle, 1988).

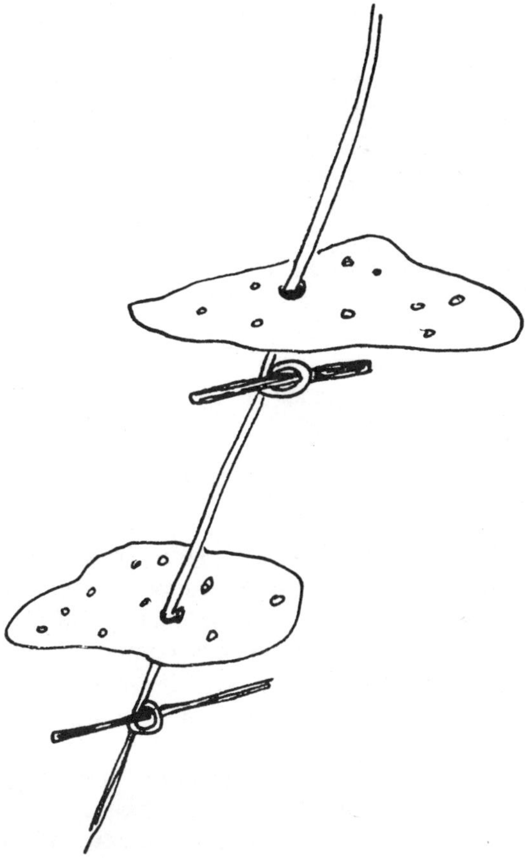

Fig. 73. A simple loop held in place by a stick can be used to separate cultch if there is not much movement of the strings or handling.

In most cases, the stringing material cannot be reused except with much effort. This means it must be expendable and, thus, inexpensive. Wire has advantages for large scale operations since it is easier to work with, but either wire or rope may be used for small scale culture. Monofilament nylon (140 lb test [63 kg]) is becoming readily available. With monofilament, a loop with a nail or a short twig inserted in it and the loop drawn tight makes a satisfactory spacer (Fig. 73).

Strings may be hung vertically from single beam racks. The tripod form is particularly useful for strings (Fig. 69). The racks should be placed at a level to evade the dense barnacle zone if there is one, and this may dictate the length of the strings. For ease in handling, strings not much longer than 1 m or so are satisfactory. Strings may also be laid horizontally on double beam racks (Fig. 70). In this case the strings may be somewhat longer. The strings are placed about 25 cm apart and care is taken to see they do not touch the bottom.

Drills or crabs will crawl up the posts so continuous surveillance is necessary. If the problem is severe, both picking and trapping programmes should be instituted. Care should be taken to be sure that the beam (horizontal pole) is always out of the water. This will increase the difficulty for crabs and drills to crawl from the bottom to the strings. (But it may not completely stop them!)

Stick Culture

Stick culture is the simplest, least costly and probably the most efficient method of oyster culture and recommended for any initial attempt to grow oysters in the tropics. It is the standard method in use in Australia and New Zealand. It is an adaptation of the natural mangrove root oyster production. The basic plan is the collection of seed on narrow sticks resembling mangrove roots. These are placed horizontally on double beam racks where the oysters grow to maturity. The original collectors in Australia were dry mangrove sticks about 3 cm in diameter and 1 m in length. Wooden sticks sawn to specific measurements are also used. Later the sticks were tarred and a further development is the use of fibre-cement slats 5 mm thick, 4 cm wide and 120 cm long. Another modification, originating in Europe, is use of plastic piping whose outer surface is finely grooved.

To collect seed the sticks are arranged in bundles (Fig. 36), with each layer separated by 2 cm spacers to allow larval penetration and tied with galvanized wire or strong twine. A bundle of fibrocement collectors measures about 30 cm x 30 cm x 120 cm. These bundles are usually placed on double racks. If the bottom is firm enough they may be stored there, but spat collecting efficiency may be reduced in comparison to racks.

After setting, the spat are allowed to grow to a length of not more than 2 cm. The bundles are then broken apart and individual sticks are placed on double beam racks (Fig. 70) at a tidal level low enough to promote optimum growth but still avoid the barnacle zone if there is one.

Wooden sticks may require nailing to the beams, with a sloping nail on each side. The fibrocement slats should be set on edge and held upright by nails. Sticks may also be made of ferrocement. These are triangular in shape, with 2.5 cm sides and usually 120 cm long. The sticks are placed 10 to 15 cm apart, depending on the market size of the particular species.

The oysters are allowed to grow to maturity on the sticks at which time they are chipped off. There will be a proportion that are less than market size, and these are allowed further growth either on trays or on suitable bottom. If the oysters are chipped off with care, the collectors may be used again.

Wooden sticks may not last very long, owing to shipworm attack unless protected. Shipworms, often of the genus *Teredo*, are molluscs which bore into wooden marine structures. Tarring sticks or other wooden structures offers some protection from marine borers. The tar may be applied hot or cold. Addition of some pitch to the hot tar gives a good surface. The proportion of pitch to tar requires experimentation for tars vary greatly in their composition. Copper compounds dissolved in kerosene and added to the tar (500 g of copper oleate in 2 litres of kerosene added to 4 litres of tar) gives added protection to wood used for the racks. Wooden trays also require protection, and synthetic fish net is preferable to metal netting for the tray bottoms. If metal netting (chicken wire) is used, it should also be protected by tar. Stick culture is subject to the same problems of predation and fouling as other intertidal culture systems.

SUSPENDED CULTURE

Also called raft or hanging culture, suspended culture includes the culture of oysters either by tray, string or stick suspended from rafts or long lines in deep water. The method with the most production potential is the string type. At one time virtually all of Japan's extensive oyster production was derived from bottom culture, but in recent years there has been a move to suspended string culture. In temperate waters, bottom culture is capable of producing 2.2 t of oyster meat/ha/yr, while suspended string culture may produce 9 t/ha/yr. This is possible partly because of utilization of the third dimension (depth) and of more rapid growth than on the bottom. The possibility of predation is also reduced, but fouling may be a significant negative factor.

Rafts

Rafts consist of a wooden or a bamboo framework supported on a flotation medium such as wood (i.e., logs), oil barrels, styrofoam or piping sealed at the ends. Wooden logs are not always available in the tropics, tend to be costly, and are subject to shipworm attack which reduces significantly the flotation capacity and the life expectancy. Styrofoam is costly and often requires its surface to be protected from abrasion. Some types are not resistant to oil slicks. Oil barrels (45 gal or 200 litres) are generally available and each barrel is capable of supporting about 450 lb or 200 kg. A four barrel raft, one on each corner, can support nearly a tonne, less the weight of the wooden or bamboo superstructure. Strings of oysters weigh about 5 times less in water as they do in air. Estimates used to design a raft or other flotation should be the expected weight at harvest. However, due to great changes in weight with oyster growth and fouling, calculations can only be very approximate.

The framework may be either square or rectangular. First a framework of bamboo or mangrove poles (A B C D) is laid out as in Fig. 74. The poles should be not less than 8 cm (preferably 10 cm) in diameter. The length of the raft will be mainly governed by the length of the poles available as well as the thickness, but if the length is too great there is a tendency for the middle of the span to sag. A length of about 6 m and a width of 4 to 5 m makes an acceptable raft which should support nearly 200 strings.

In Fig. 74 the poles at the corners are allowed to protrude. Where they cross, a 15 to 20 mm hole is drilled through both poles and a peg of the same diameter is driven through. Joints are then tied with several crossing loops of wire, twine or 5 mm diameter rope. The frame is then lifted and supported on barrels placed in the corners with the top protruding 15 to 20 cm above the

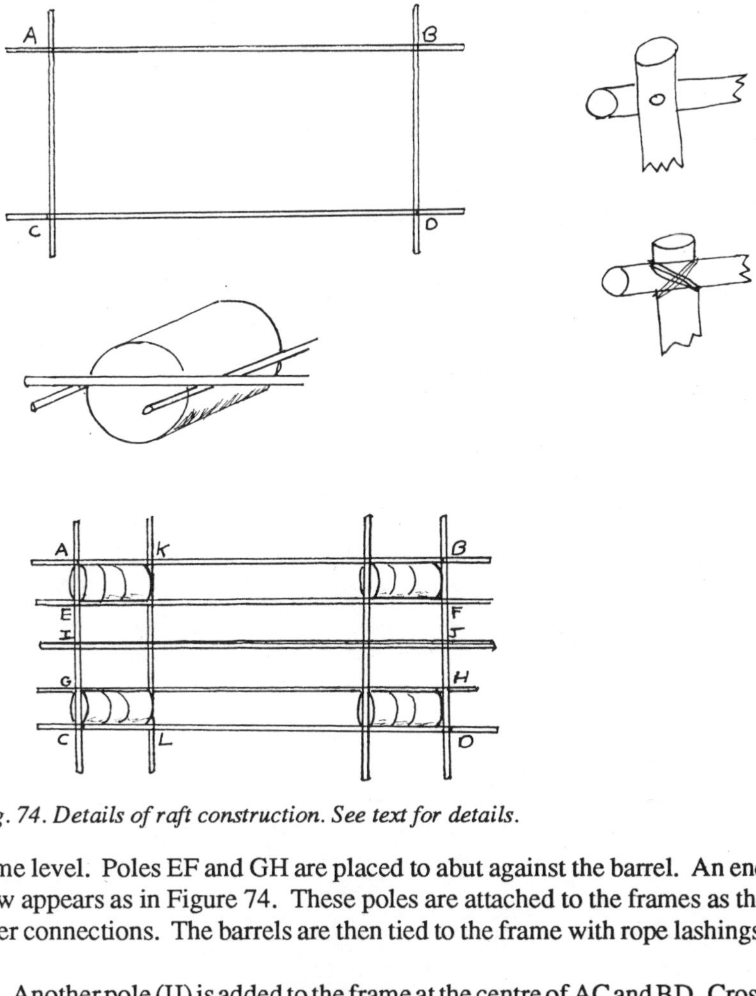

Fig. 74. Details of raft construction. See text for details.

frame level. Poles EF and GH are placed to abut against the barrel. An end view appears as in Figure 74. These poles are attached to the frames as the other connections. The barrels are then tied to the frame with rope lashings.

Another pole (IJ) is added to the frame at the centre of AC and BD. Cross members such as KL are placed on the frame at 1–m intervals and tied at the intersections with AB, IJ and CD. Pegs can be inserted at each intersection but the main positions for pegs is at the corners. The frame should now be quite rigid but if there is any indication of an accordion motion, a diagonal piece from point B to point C may be attached.

The oil barrels should be tarred to prevent corrosion. If the tar locally available tends to be sticky, a small amount of pitch may be melted into it. An alternative to tar is to cover the drums with ferrocement, the procedure for which is explained in Appendix III. If fouling is severe, it is possible to cover the barrels with plastic sheets or large bags which can be periodically changed.

Long Lines

An alternative to rafts for flotation are long lines which are long ropes or cables anchored at each end and supported at intervals by small floats (Fig. 75). The size of long line components depends on the length of the long line and the amount of wave action to which it is subject. The main line rope should be, however, not less than 12 mm in diameter and for long life should be synthetic rather than manila, which has a relatively short life. Strings should be hung on the line at about 3 or 4/m. Depending on the total number of strings to be exposed, the length of the line may be calculated. Long lines of more than 100 m are often difficult to manage. More lines of moderate length (50 – 100 m) are better than a few very long ones. A thimble should

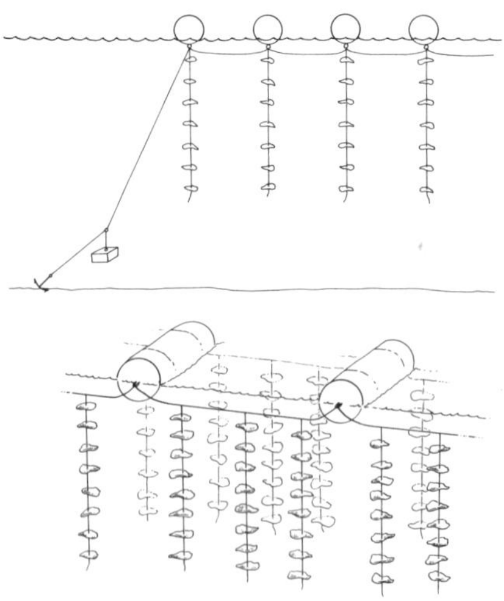

Fig. 75. a) Single float surface long line with weight for keeping the line taut and damping wave action. b) Double long line. Surface type (from Quayle, 1988).

be spliced into each end of the line for attachment of the anchor line or lines to prevent wear by abrasion. The end of each of the two anchor lines should also be thimbled and attached to the main line by shackles (Fig. 76) whose pins should be secured to the shackle by a wire. This is to prevent the pin becoming loose which it is apt to do from constant tugging. Anchors and anchoring methods are explained later.

There are many types of long–line floats, and the major deciding factor is cost. There are plastic or styrofoam floats specifically designed for long lines, but any of a wide variety of fishing floats, such as Japanese glass balls, may be used. Styrofoam logs can be cut up into single small floats. Various types of discarded industrial barrels or bottles are suitable and often available relatively cheaply and among these are oil barrels. Discarded automobile

Fig. 76. (left) A thimble with a rope end spliced around. (right) A shackle.

tires filled with styrofoam or urethane foam are also used. In countries where logs are available and cheap, small sections may be used. In the tropics where bamboo is generally available, a number of short sections about 1 m in length can be bound together in a bundle to make a float. An alternative type of long line with bamboo is to make long (6 to 10 m) bundles of about 50 cm in diameter. These can be tied end to end and strings suspended directly from the bundles. Ferrocement makes good floats. They are easy to construct and are not subject to attack by boring organisms. Ferrocement can be used to cover barrels, styrofoam or anything which can act as a mould. The procedures for using ferrocement are given in Appendix III.

Construction should be carried out ashore by first laying out the main line and splicing the thimbles into the ends. Anchor lines are attached after these have been thimbled at each end. Anchor lines should be about 3 to 4 times as long as the depth of water at high tide into which the long line is to be set. The floats may then be attached, the method depending on the type of float. The main line may be attached as closely as possible to the floats or it may be dropped to whatever distance below the surface is desired by means of short drop lines. Attaching strings or trays and harvesting is accomplished more readily with the line close to the surface.

An oil barrel, owing to the relatively large weight it can carry, can hold a double long line. With a rope not less than the diameter of the main line, a tied loop (Fig. 77) is made in the middle of a length of line which is then looped at least once around each end of the barrel and tied. A short line joins the two barrel loops together so they will not slip off the barrel even if a little loose. The main line is then attached to the barrel loops by lashing with strong fishing line. Assuming a string of full–grown oysters (7 to 8 cm) weighs 5 lb (2.3 kg) in water then one barrel can support approximately 50 strings. At 6 strings/m (double line), barrels may be placed on the main line at about 8– to 10– m intervals. This should be used as an approximation and adjustments may be made as experience is developed.

Bundles of 1–m long bamboo may be harnessed in a manner similar to the oil barrels. Experience as a result of trial and error either with a known weight or with actual shell strings will determine the carrying capacity of bamboo bundles in a long line.

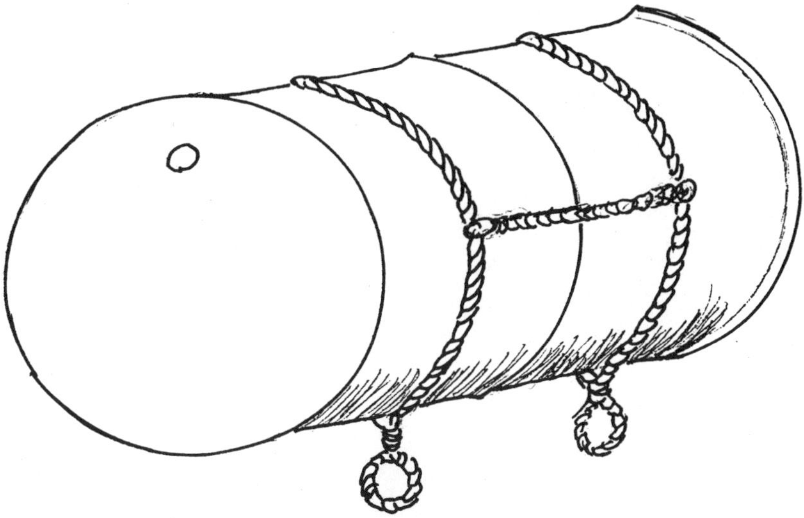

Fig. 77. A barrel with a harness to be used to float a double long line.

Anchors and Anchoring

Rafts and long lines require means to hold them in place. They may be held in place by tying them to permanent or temporary pilings or by lines to shore anchors in sites where there is no boat traffic. However, the most important method is by anchors, of which there are many types (Fig. 78). The most efficient anchors do not depend solely on weight. They use either suction forces (in mud) or dig into the substrate. Most of these are manufactured but some, like the mushroom anchor (Fig. 79), can be readily made from cement on site. In addition to the cupped form of mushroom anchors, which relies on the edge digging into the bottom, there is also an inverted mushroom type which relies on suction and is useful on muddy bottoms (Fig. 79).

Of the manufactured types, the plow anchor is considered to be quite efficient, but the Danforth has good holding power relative to its light weight. For instance a 2–kg Danforth is considered to have the same holding power as the standard 45–kg mushroom. A 6 x 3 m raft with strings requires about a 9–kg Danforth or a mushroom type of a few hundred kilograms.

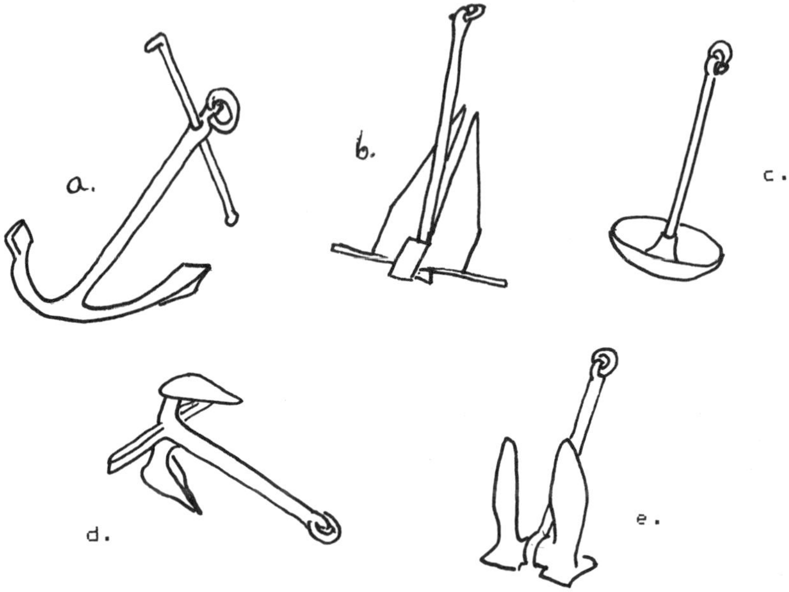

Fig. 78. Five types of anchors. a) Yachtman's, b) Danforth, c) Mushroom, d) Northill and e) Navy.

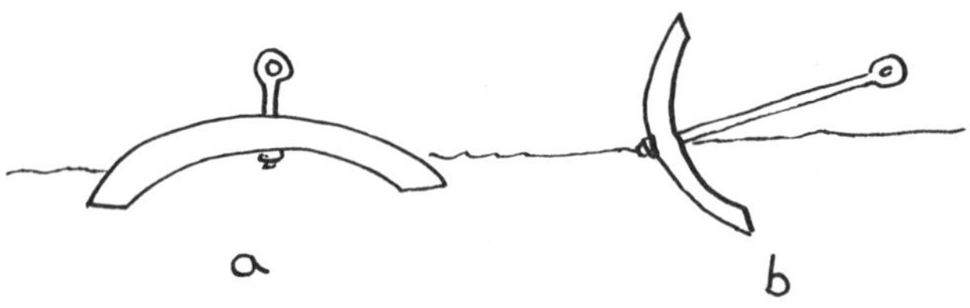

Fig. 79. a) Inverted mushroom anchor and b) cupped form of mushroom anchor.

The main advantage of the mushroom anchors is the low cost and availability. To make an ordinary mushroom (drag) type anchor, a form, e.g., a plastic wash bowl, of the desired size is needed. Also a hole of the required shape dug in the sand and lined with plastic can serve as a form (Fig. 80). The forms are filled to the desired thickness with a stiff cement mixture; if it is too thin, it is difficult to obtain the moulded shape. A suitable cement mix would be the proportion of cement to sand to gravel of 1:2:3 by volume. A standard sack of cement weighs 36 kg and is 27 l in volume.

In making the anchor, a hole through the centre is required or the holding stock, either metal or rope, is inserted before the cement is poured. The latter is particularly necessary in the case of the suction anchor which relies on water-tight suction.

About 5 days is required for full curing. In the high temperatures of the tropics the curing cement should be kept moist by light sprinkling of water.

Other anchors are rocks tied with chain, old car engine blocks, cement–filled 15–litre oil cans or oil drums as well as native fishermen's inventions. When cement filled cans are used two or more iron bars inserted through the can and protruding on either side will enhance the drag and thus the anchor's efficiency.

Anchor Lines

Anchor lines, along with their points of attachment to the anchor and raft, are of most importance, for it is usually here that trouble occurs with raft moorage and seldom with the anchor itself. The tendency is to use rope of too small a diameter and most often this turns out to be false economy. For a 6 x 3 m raft the anchor line should be 20 mm (3/4 in.) in diameter. At the anchor end of the line, especially if it has a metal stock, a thimble should be spliced into the end of the rope line. This end of the line is then attached to the anchor with a shackle whose screw pin should be wired fast (Fig. 81). Preferably between the anchor and the rope there should be at least 4

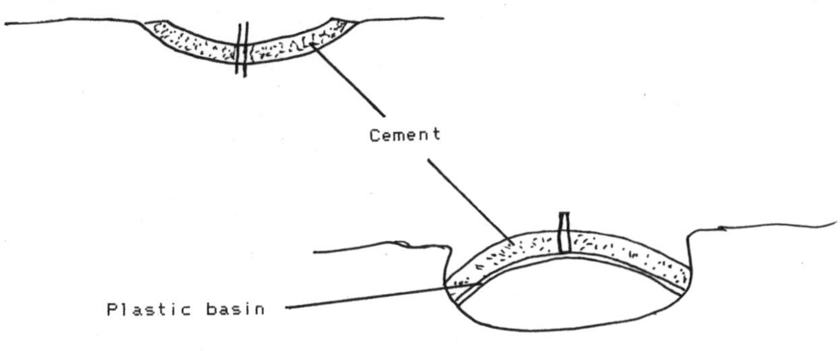

Fig. 80. Two ways of making a mushroom anchor.

m (2 fathoms) of 10 mm (3/8 in.) diameter chain for a 6 x 3 m raft in 12 m (6 fathoms) of water. The weight of the chain acts as a buffer to prevent jerking of the raft from wave action. Attachment to the raft, if on wood, should be with a double loop of rope with a bowline knot. A rule of thumb for mooring lines is "metal to metal and rope to rope." A rope tied directly to an anchor will chafe and eventually break. Another rule of thumb is for the length of the anchor line to be about three times the depth of water.

Anchoring

The simplest form of anchoring is where there is a single anchor at each end of a single raft or long line. Two anchors are required; otherwise, the suspended material will become entangled with the anchor line.

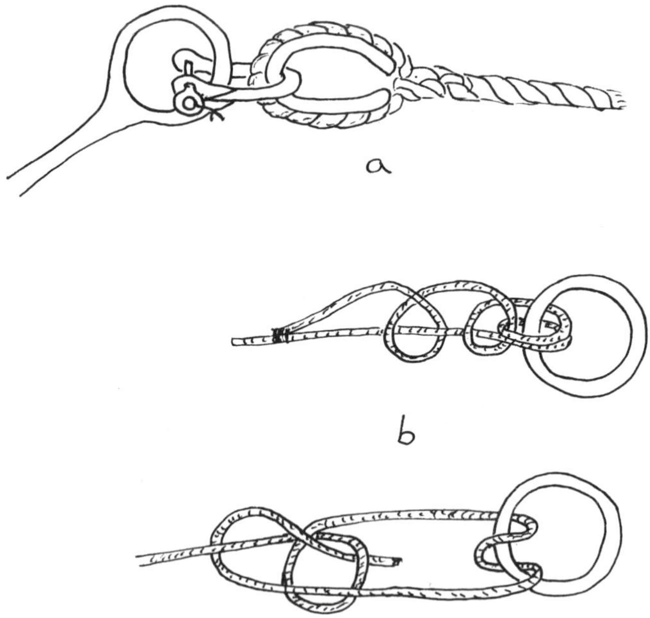

Fig. 81. a) Anchor line is secured to the anchor with a shackle (note wire on pin) between thimble and anchor eye (this could be an anchor chain). b) A fisherman's bend (above) or an anchor bowline (below) are used to secure the anchor line to a raft.

In setting out a raft or a long line, it is essential to be sure that the system is tight. With a two anchor system, the anchor at one end of the raft or long line is dropped to the bottom after making sure the anchor line is firmly attached to the raft or longline. Through the eye of the stock of the other anchor a loop of light rope is passed. The length of the loop should be at least double the depth of water and the two ends of the loop secured to the work boat (Fig. 82). This anchor is now lowered from the boat until it nearly reaches bottom. The boat is then given full power to stretch both anchor lines as far as possible. When this point is reached one end of the loop on the towed anchor is released (cut) and the anchor drops to the bottom with its line fully extended. This operation should be conducted at low water slack.

If there is little tidal rise and fall, the anchor lines will remain fairly taut if they have been set properly. If there is a considerable tidal range, a heavy weight suspended on the anchor line at a point where it will not touch bottom (about halfway) will maintain tension on the line (Fig.75). This is particularly important for long lines to prevent tangling. Experimentation will be needed to determine how much weight is needed, since this depends on the size of the long line and the amount of flotation.

Alternate Anchoring Systems

If there are two or more rafts in tandem, they should be secured to each other at the corners (Fig. 83a). If there is a strong tidal stream, a bridle for each anchor line may be useful (Fig.83b) . In difficult anchoring situations, it may be necessary to use double anchors (Fig. 83c and d).

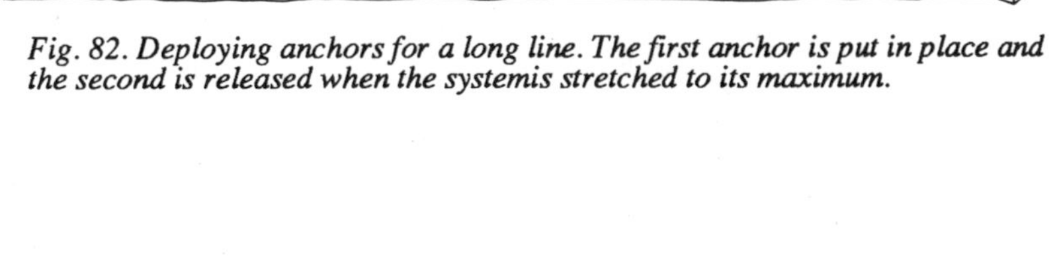

Fig. 82. Deploying anchors for a long line. The first anchor is put in place and the second is released when the systemis stretched to its maximum.

String Culture

The first move from bottom culture was the use of strings on rafts. As this method expanded into more exposed waters, long lines were used as they are better capable to cope with greater wave action, as well as being less costly. String construction has already been described. The actual method to be employed depends on the type of cultch, type of flotation and availability of materials. Shell cultch will sink so it does not need weights. However, with some tropical oysters the shells are thin and light and, although they will sink, the slightest current moves the strings enough to cause entanglement. In this case, as with bamboo cultch, it is necessary to attach a small weight such as a stone to the end of the string.

The length of the string is partly a function of the depth of the water and partly of the manpower required to lift the strings. Strings should not touch bottom at the lowest spring tides, because of predation or silting. Caution is needed to prevent the swinging of the raft or longline into shallower water. With a spacing of 30 cm around 10–15 pieces of cultch, the strings are manageable while providing reasonable production. Strings up to 10 m in length are used in mechanized operations. One problem in lifting the strings

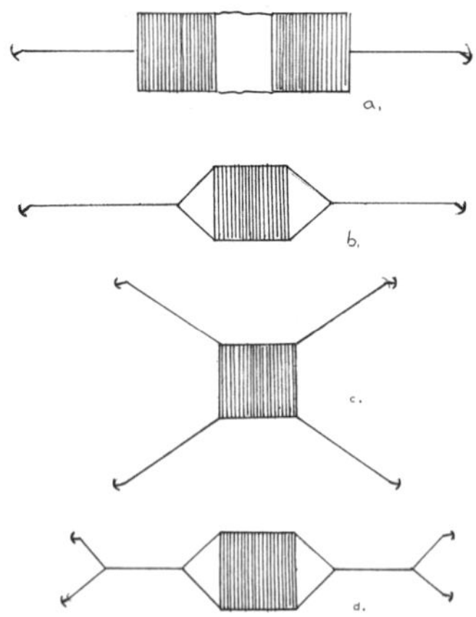

Fig. 83. Alternative anchoring systems. a) Rafts in tandem can use the same anchors if they are large enough. b) A strong tidal flow may require a bridle. c) and d) Difficult circumstances may require multiple anchors.

at harvest is loss of oysters as the clumps strike the flotation gear or the side of the harvesting vessel. This can and should be prevented.

In the tropics there is seldom thermal stratification but there may be a halocline and in some cases, in some seasons, it may be necessary to place the strings below the halocline to reach higher salinity water. This will govern the length of strings.

The typical procedure is the collection of seed on strings with tightly spaced cultch with a single string of 100 pieces or more. When the time comes to place strings on floats or racks the cultch are removed and restrung at 30–cm intervals for growing. This is usual when seed areas are distant from the growing areas. This may also occur when the seed collecting area and growing area are at the same site. The alternative is to string unseeded cultch at 30–cm intervals and collect seed so restringing for growing is not necessary. Which method to use depends on growth rates, fouling sequence, setting patterns, labour costs and timing of other fishing activities. Initial trials with both methods may be necessary.

Spacing

Strings are hung on rafts or long lines spaced about 50 cm apart. This allows adequate water flow between them and provides enough space for growth so clusters on adjacent strings do not touch. Rope strings are attached with a clove hitch, while wire strings should be given two turns round the rope or raft before securing with two or three twists.

Cultch

Various types of cultch for stringing have already been described. Of importance is the number of spat on a piece of shell cultch for optimum production. Mortality of string–cultured oysters emanates mainly from competition for space on the cultch. An optimum number of about 15 – 20 market oysters of a length of 8 – 10 cm may be expected on a 10–cm oyster shell cultch. The minimum number of spat required for this is 25 – 30 spat/100 cm^2. A smaller number of spat will produce a greater percentage survival but not a greater absolute number of market size oysters. Excessive numbers of spat on the cultch piece (i.e., more than 100) are taken care of by competition for space so it is not necessary to attempt to reduce the number. The minimum number for an economic return is critical and this is probably 10 to 15 spat/100 cm^2. Small numbers will give a high percentage survival but there is also the risk of producing non-cupped flat oysters due to xenomorphism. Growth rate of smaller numbers is somewhat better but not enough to warrant deliberate evasion of large numbers.

Since in suspended culture the flotation system is the most costly item, it is necessary to utilize it to the optimum. This means keeping the oysters there for as brief a period as possible. One way of doing this is to string cultch with relatively large spat providing the cost to produce the large seed is less than the cost of putting the seed directly on the long line. In temperate waters it is advantageous to hold the seed for 1 year in bottom culture until it reaches a size of about 30 mm before floating. This enables it to reach market size with one summer's growth in suspension. This size of 30 mm permits the oyster to better withstand the effects of fouling and is considered to reduce the mortality and improve meat production. In the tropics, with the brief period required to produce a marketable oyster, this preflotation period may not be necessary in all instances, but should be considered.

Productivity

Productivity will vary according to the site, growth rate and size of oysters cultured. Typical yields in temperate waters will be presented here. Similar yields can be expected in most situations, although the time to harvest will vary greatly. A 12 m x 3 m raft can support about 100 strings of 15 pieces of cultch and 1 ha should be able to hold 50 to 60 rafts (25% of the area) or about 2500 m of long line (2 strings/m) providing one string/2 to 3 m^2 of water. Assuming only 10 oysters/2 to 3 m^2 cluster or 150 oysters/string are produced, the yield will be about 25 kg of oysters in the shell or 3.5 litres of oyster meat/string. (One litre of oyster meat weighs 0.80 kg.) Thus, a raft or a long line of 50 m holding 100 strings will produce 2500 kg of whole oysters or 350 litres of meat, and with 50 long lines/ha the yield will be 125 t whole oysters. These calculations are only approximations and based on oysters about 10 cm in length.

Floating Stick Culture

Collection of seed for floating stick or tube culture is not common but is done in exactly the same way as for stick culture on racks. The routine of separating bundles of sticks when the spat reaches the appropriate size is also the same. The next step is arranging the suspension of sticks from the raft or long line. This may be accomplished by arranging the sticks horizontally in parallel one below the other and suspending with a bridle (Fig. 84). The spacing of the sticks should be about 30 cm apart. The length of the tier will in part be controlled by available means of harvest. If harvesting by hand, more than approximately 5 sticks will be difficult to lift without losing oysters by breakage from the clusters. If harvesting by an A–frame, davit or crane, the number is limited only by water depth or height of the lifting device.

Alternatively, the sticks can be predrilled with a 5 to 7 mm hole at each end (Fig. 84), so they may be suspended vertically, either singly or end to end. They may be attached about 40 cm apart to a raft or a long line with cord, wire or nylon. Removal of mature oysters from the sticks is similar to that practiced with rack stick culture. Bending or flexing of plastic tubes removes most of the oysters.

Tray Culture

Trays may be used in floating culture as well as on intertidal racks. Tray construction is similar to that described previously but the trays are held in stacks. Manufactured trays have special apertures through which rope may be threaded from one tray to another to hold them together. Hand–made trays may be made with similar provisions or they may be held together in a bridle (Fig. 85). From the labour standpoint, large trays are usually more economical to build than small trays. For artisanal culture, trays should be small enough to permit lifting without mechanical aid, so a compromise is necessary between few large trays or many small trays.

Mesh may be made of manufactured metal (chicken wire), plastic mesh or fishing net. If fishing net is used, centre cross bars are required, for it tends to sag in the centre where the oysters will congregate. Mesh size is a

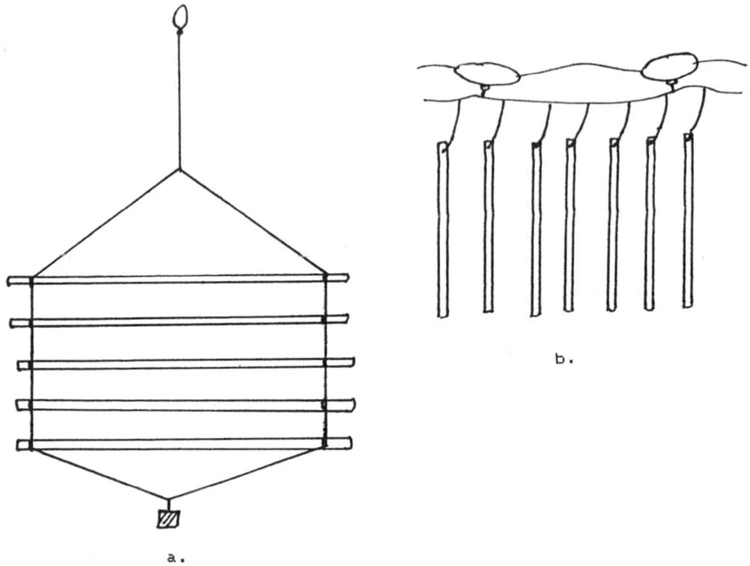

Fig. 84. a) Bridle for sticks. b) Sticks hung singly from a long line.

compromise between oyster size and need for adequate water flow through the tray. Different trays for different oyster sizes or provision of finer mesh liners for the smaller oysters may be required. Other types of trays are the pearl nets and lantern nets, either manufactured or hand made (Fig. 86).

Owing to the high cost of trays, it is necessary to make optimum use of the tray area. One more compromise is that between reduced growth from crowding and the use of more trays with less crowding. A factor in this is the seasonal sequence of growth and of marketing. In most instances, fullest use of a minimum number of trays required for the production target is indicated, but the economics must be determined for every case considering the number of trays and the turnover.

A serious problem with tray culture is fouling. Oysters within the trays, particularly with a cover, are seldom seriously affected by fouling. The major difficulty occurs on the outside of the tray where the pores can be covered over, thus reducing water flow. Keeping a free water flow is most important so frequent cleaning is necessary. Fouling on suspended trays is

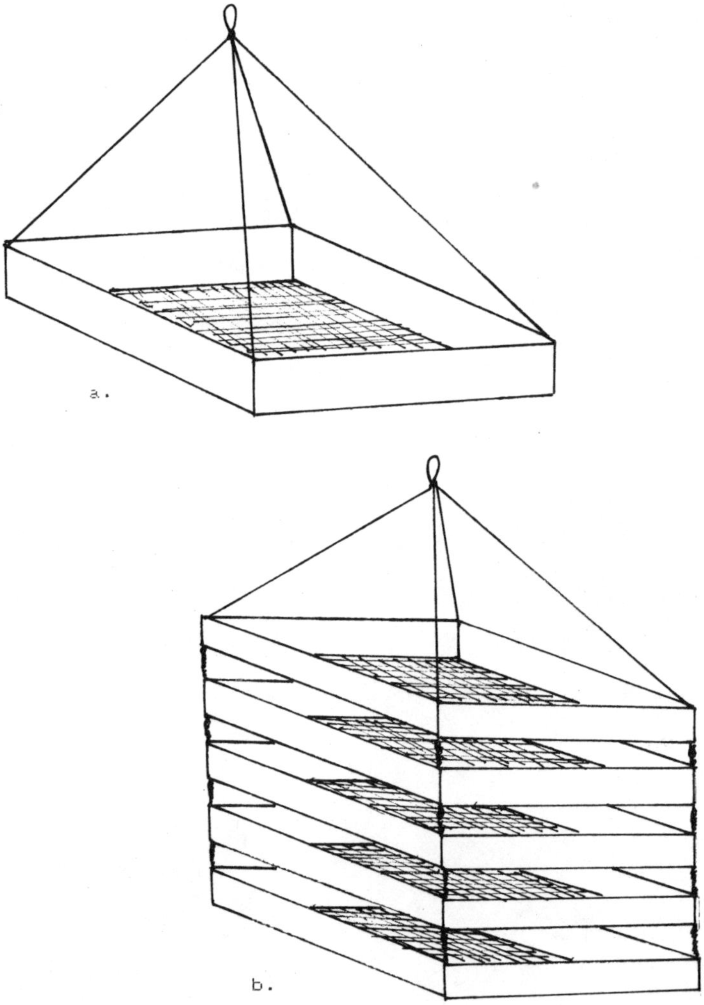

Fig. 85. Trays constructed of a wooden frame with a mesh bottom. a) Single tray held by bridle. b) Multiple trays.

invariably greater than on intertidal trays. If the fouling sequence with season and depth is known, it may be possible to manipulate around it, either by holding the trays intertidally during certain periods of heavy fouling or by placing the trays at a different depth.

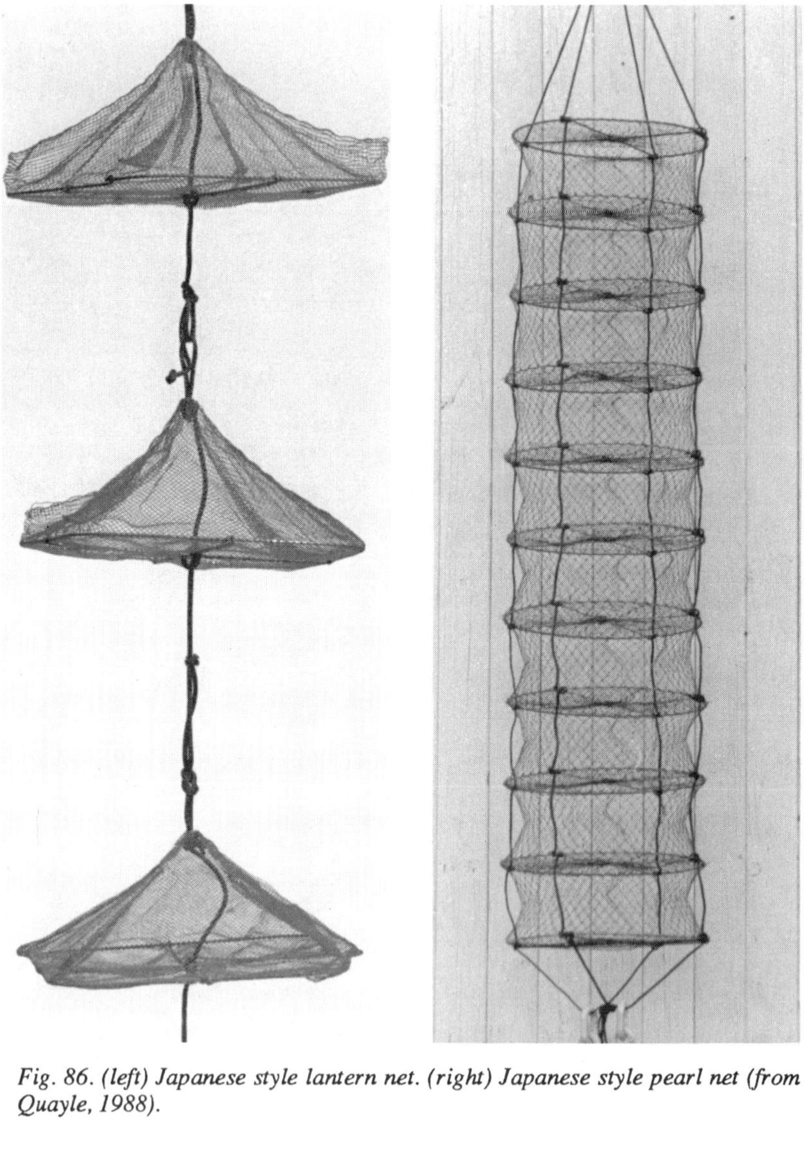

Fig. 86. (left) Japanese style lantern net. (right) Japanese style pearl net (from Quayle, 1988).

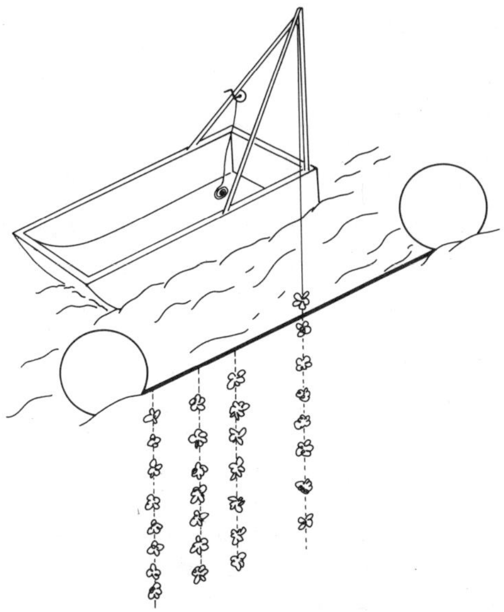

Fig. 87. A–frame for harvesting.

In general tray culture presents many problems and is usually more costly than most other types of culture and should not be attempted except as a last resort (Table 7). If the purpose of the culture is to provide a market with half shell oysters then tray culture will do this. However, string culture and bottom culture will provide a proportion of oysters satisfactory for the half shell trade while also producing shucking oysters at a much lower cost than tray culture.

Positive factors	Negative factors
Rapid growth rate	High cost of trays
Single oysters	Suspension costs
High quality meats	Fouling
Good condition factor	Excessive labour
Good shape	Questionable Economics
Control of Stock	Single seed required

Table 7. Positive and negative aspects of tray culture.

Harvesting Suspended Culture Oysters

Strings may be pulled up by hand, taking care not to touch the side of the boat or raft or long line because some of the oysters in the clusters will be only lightly attached and easily dislodged. A simple A– frame or davit with a rope and pully may be placed on a harvesting raft or boat (Fig. 87). If there is a sufficient tidal range, the raft or long line may be towed into the intertidal area at high tide and the strings dropped to the bottom where they may be picked up at low tide.

Clusters are removed from strings simply by clipping the wire or cord between them. If the clusters are separated by tubing, the strings are best suspended to allow the clusters to slide down. Harvesting is the time to wash mud and silt from the clusters and remove most of the fouling.

Separating oysters on clusters into singles or doubles is done before they are shucked, as this process interferes with shucking, in which the development and maintenance of a rhythm aids efficiency. Keeping cluster separation apart from shucking also allows selection and immediate planting of undersized and shell–damaged oysters on racks or on the bottom where they will be grown to market size. At this time, oysters suitable for the half shell market may be separated from shucking oysters.

Since the shell of suspended cultured oysters is soft and thin, shucking in the normal manner from the side may be difficult because of shell breakage at the adductor muscle. Instead, the oyster knife is sharpened to a point and jabbed into the bill (posterior end) of the upper valve, just short of the adductor muscle and opened from there.

PROCESSING

Oysters are sold either in the shell, termed "half shell," or as shucked meats. Both methods are practised in the tropics. Since refrigeration is a problem, it appears that the half shell method should be preferred, for oysters store reasonably well in the shell as compared to the shucked condition.

Shucking

Oysters should always be opened with gloved hands on a firm table and not held in the left hand. The shell is placed on the table with the umbo to the left with the cupped valve down. The adductor muscle is found about two thirds of the distance from the umbo where the oyster knife is inserted and with a twist, the muscle is severed. The two valves are then pried apart and the upper (right) valve lifted off. The muscle attachment on the left or lower valve is cut with the knife and the oyster meat is lifted out.

Oysters, particularly small ones, may be opened by prying open the hinge. There is also another method whereby a pointed knife is jabbed into the posterior end of the oyster, the so-called bill, and the oyster opened from there. There have been many attempts to open oysters mechanically but so far these attempts have been unsuccessful. Many chemical methods have also been attempted but, other than causing the shells to gape slightly, little success has been attained.

Heat has also been attempted with moderate success and apparently microwaves will open oysters but so far no industrial application has been found. In Africa, oysters on mangrove roots are steamed until they open and then the meats are picked out. This probably provides some pasteurization but there are no controls.

There is a heat shock method where the oysters are given a brief dip of 2 – 3 minutes in water at a temperature of 140 – 150° F (60 – 65° C.). This causes relaxation of the adductor muscle without cooking the oyster meat and is followed by an immediate chill. Again all the normal shucking motions are still required. This system may be useful for *Saccostrea cucullata* which is notoriously difficult to shuck. A high level of sanitary control is required for this method.

Washing

After the oyster is shucked, it should be washed. Either salt or fresh water is satisfactory as long as it is bacteriologically pure. In a large–scale operation, air is bubbled from the bottom of the container in which the oysters are held. This washes without bruising. Smaller operations normally use deep stainless steel sinks with a colander insert and stir the oysters gently by hand or with paddles. Rubber gloves should be worn when handling raw oyster meats. After washing, the oysters are poured into shallow perforated stainless steel trays. Here, the oysters receive a final wash from an overhead hand held spray and it is here the oysters are graded if necessary.

Sanitary regulations in most countries require that if shucked oysters are washed in fresh water, the period should not exceed 3 minutes. Any longer than three minutes cause the oysters to absorb water and lose body fluids. This process is called plumping and is against most regulations.

Packaging

Basic sanitation regulation regarding packaging states that shucked shellfish shall be shipped and sold, retail or otherwise, either in single service containers made of clean impervious materials positively sealed or in such containers so sealed that tampering with the container can be detected. After sealing the containers must be airtight. This is done to ensure a minimum amount of handling, for each additional step increases the hazard of contamination.

Containers are also marked so it is possible to trace a particular shipment to its origin. These are often marked with the date of packing to give an indication of how fresh the contents may be. Containers are sized to conform to the market for which they are intended, and are of waxed paper, plastic or glass.

Storage

The storage of shell stock (oysters live and in the shell) has already been described. For retail purposes, shellstock should be held in cold rooms or stored with ice. There should be no contact of the oysters with ice in case the ice is not bacteriologically safe. Oysters store better in a dry condition. The shelf life will vary with the species, some being better adapted to not gapping during storage.

Storage of shucked oysters is much more difficult. They should have immediate refrigeration after packing. In the packed form it takes some time for the oysters to attain correct storage temperature, so every effort should be made to speed up this process. The larger the container, the longer is the refrigeration time to reach a satisfactory storage temperature. Tests have shown that oysters held at 12° C were unacceptable after 3 to 5 days, while those held at 8° C were unacceptable after 7 to 8 days, and those held at just above 0° C (1 - 3° C) were still satisfactory after 16 days. The importance of good refreigeration cannot be overemphasized.

Processed oyster products

In addition to the standard fresh and half shell oysters, there are several ways of processing oysters: smoking, canned whole oysters, canned oyster stew, freezing, making oyster soup and sauce.

Smoked oysters

This method requires shucked oysters that are partially cooked or have been steamed open. They are rinsed for 5 minutes in a 2.5% brine solution to salt the product slightly and to wash off sand or grit. They are then spread in a single layer on a 1 cm mesh wire tray which has been previously brushed with cooking oil to prevent adherence of the oysters to the wire. To prevent oxidation and discolouration of the oysters, the smoking process should be ready to receive them. The smokehouse smoke should have a temperature of 45° C. Smoking time is variable but about 4 h at this temperature gives a light brown colouring without causing the meat to become shrivelled or tough. The type of wood used and the time and temperature is largely a matter of individual taste and opinion. Smoked oysters have a longer shelf life than unsmoked oysters and may be canned or sold as is, although they should be refrigerated.

Canned oysters

Oysters may be canned whole in brine after shucking, either by hand or by steam. They should be blanched before being canned in the standard manner.

Oyster stew

Oyster stew forms an outlet for oysters which may be too large, or for those that do not meet normal market standards, were badly cut in the shucking process or low in condition. They are prepared as for canning but are chopped into small pieces and added to milk and spices.

Frozen oysters

Freezing is an excellent way of preserving oysters providing refrigeration facilities are available. The oysters should be blanched before being individually quick frozen, after which they may be stored for a considerable period of time without deterioration.

Oyster sauce and soup

Oyster sauce is an important item in Oriental cooking and should provide a use for substandard shucked oysters. Such oysters are also useful for prepared soups. Local processors should be consulted regarding the preparation and potential marketing of these products.

CHAPTER 7

MUSSEL CULTURE

Mussels are among the most familiar of all bivalves for they are widely distributed throughout all oceans. They occur both intertidally and subtidally, often in tremendous numbers on both rocky and soft bottoms. While generally immobile, they have the ability to move short distances.

In countries where natural populations exist, mussels are seldom cultured if not often a part of the normal diet. The main centre of mussel culture and utilization has been western Europe. Production in the tropics has not been extensive, but there is increasing interest and activity in mussel culture. It is also expanding rapidly in North America. Chile and Peru have always had extensive mussel fisheries. Where attempts have been made to establish mussel culture, the main problem has been markets rather than technical or biological problems. The meat of mussels is equal in food value to other molluscs and should provide a reasonably inexpensive source of protein.

Mussel culture techniques are not too different from those for oyster culture. Most techniques are based on the ability of mussels to attach with the bysuss. Bottom culture is practiced in the Netherlands and other areas of eastern Europe. Suspended culture from rafts is used mainly in Spain. Long lines are also used for suspended culture in several countries. France has an intertidal stake system with rather large stakes (bouchots), and a subtidal stake system with bamboo poles is the method in the Philippines. In Europe the main species is *Mytilus edulis* or *M. galloprovincialis*; in the Philippines it is *Perna viridis*.

(blue mussel)

COLLECTORS

Mussel collectors take a variety of forms, depending partly on the type of culture. In the Spanish system, the major collector is synthetic rope about 20 mm in diameter, discarded from fishing operations where it has been frayed from use. These ropes are suspended vertically or horizontally from rafts. In eastern Canada, plastic netting (Vexar) has been used as a collector as well as for growing. In the Philippines and Thailand, bamboo poles are the standard collector and growth substrate. While mussel larvae have preferences, they are not highly selective of the setting surface as long as there can be a byssal attachment. This excludes surfaces with mud or silt.

SPANISH SYSTEM

Mussel culture is practiced in five inlets (rias) on the northwest coast of Spain. From large (20 m x 20 m) rafts are hung ropes (20 mm diameter) discarded from fishing operations. These may be as long as 10 m. At intervals of about 40 cm sticks (dowels) 30 cm x 2 cm by 2 cm are placed between the strands of the rope. These prevent mussels from sliding down the rope when growth has increased the weight to such an extent that byssal attachments are insufficient to hold the clusters in place. Seed is obtained either from rocks on the shore or from ropes hung from rafts. The 10– mm seed scraped from the rocks is wrapped onto ropes by fine rayon netting which rots away after several days, by which time the mussels have become byssally attached to the rope and to each other. Ropes used for seed collection are removed from the water when the mussels reach a length of about 10 mm. When the original setting density is too great for adequate growth and survival, mussel seed are removed from one rope and restrung on three or more ropes .

Market size of 8 to 10 cm require about 18 months from setting, and a 10–m rope may produce over 100 kg of mussels in the shell. The standard 20 x 20 m raft can hold 500 or more ropes, so a single raft may produce over 50 t of mussels in the shell. Meat yield varies between 30 to 50% of the whole weight. There may be 6 rafts/ha.

Harvesting is accomplished from vessels fitted with winches for lifting the ropes. Prior to the lift, a metal mesh basket 2 m in diameter is lowered and placed below the rope of mussels to be lifted to catch mussels that may slide off the rope as it rises above the surface of the water. Under–sized mussels are retained to be wrapped onto new ropes, while those of market size are sent to purification plants.

Predation occurs with fish and in some places with ducks. The main fouling organism in Spain is the tunicate (ascidian) *Ciona*. The costly rafts used in Spain may be replaced by long lines, usually the case in new mussel culture developments.

As a result of pollution, all mussels grown and harvested in Spain are artificially purified by holding them in baskets in tanks of chlorine–treated sea water for at least 48 h. These purification plants are large, privately owned and centrally placed and may process mussels from a number of growers. Each plant employs a bacteriologist, with occasional checks by government inspectors.

A proportion of the production is used fresh for local consumption, a proportion is canned and some is exported fresh. Spanish mussel production is in excess of 50,000 t annually (in the shell).

BOTTOM CULTURE

Most bottom culture of mussels occurs in northwestern Europe in the North Sea area and is essentially a subtidal operation. Seed is dredged from natural populations when they are about 10 – 15 mm in length and transplanted to the farmers' leased beds at average depths of 5 m. This ground is usually quite muddy, and before marketing the mussels are transplanted again to a firm sandy ground.

This is a highly mechanized culture with fairly large dredge vessels and with machines for washing, separating clusters and removing the byssus. This is not to say, however, that bottom culture cannot be carried out on a smaller scale on intertidal grounds or without mechanization.

STAKE CULTURE

The origin of stake culture was in France which produces 50,000 t of mussels annually. The stakes (posts or poles), called "bouchots" in France, are 20 cm in diameter and 3 m in length and are driven into the bottom to a depth of 1 m in the intertidal zone. The poles are spaced 1 m apart in separate rows 3 m apart. There are 1100 km of bouchots along the French coast. Seed is collected on horizontal ropes suspended between poles in the south of France and transported to northern areas. Here the seed is placed in plastic net tubing (Netlon) and wrapped spirally around the bouchots (Fig. 88). Some mussels migrate from the tubes and attach to the wooden bouchots. In some cases, thinning may be required later. The mussels are harvested in 12 to 18 months when 6 to 8 cm in length. This method is the culmination of 700 years of development and is suited to the particular environmental situation in France, where large tides up to 15 m expose extensive areas of mud and sand flats.

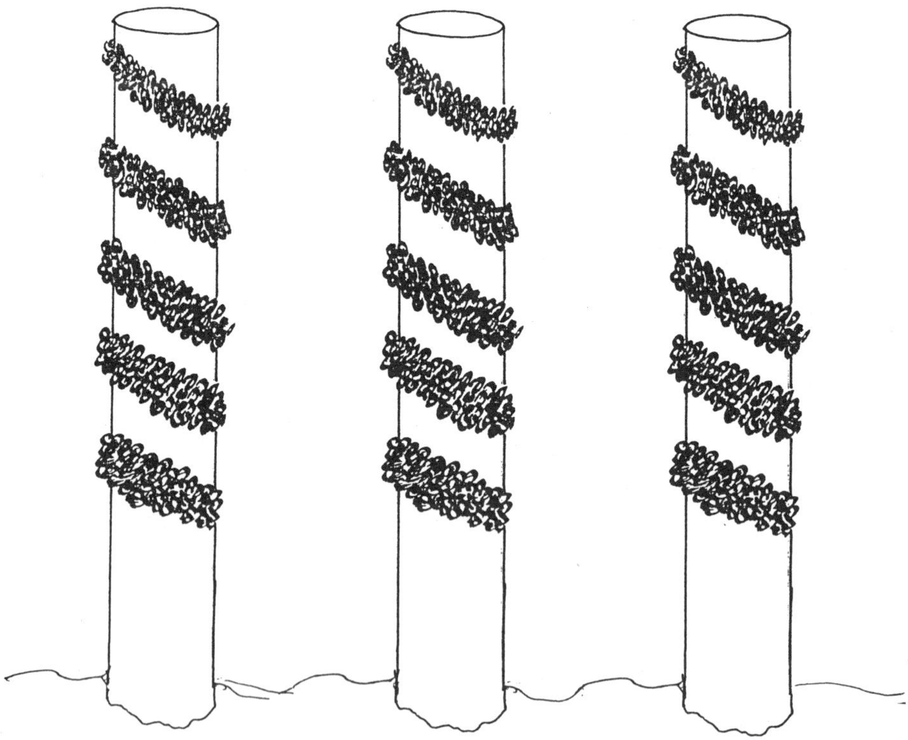

Fig. 88. Bouchots used for culturing Mytilus edulis.

All of the French production is used in the fresh state. Since most of the culture is carried out in almost open ocean conditions, sanitation is not a significant problem.

Stake culture, usually in the form of long bamboo poles is carried out in Southeast Asia, particularly in the Philippines and in Thailand. Bamboo poles are driven into the bottom in water with a depth of 2 to 10 m and spaced 1 to 2 m apart in rows. In some instances the single poles are supported by horizontal poles. Alternatively up to 10 poles are staked in a 2 m circle and tied together at the top to form a wigwam. This structure is able to withstand wave action better than single poles.

Setting takes place on the poles and up to 200 spat/m^2, on an 8 cm diameter pole, are obtained. Obtaining a barnacle set on the poles first will enhance the setting of mussel spat. Spat are allowed to grow to market maturity of approximately 50 mm in length. This requires 6 to 8 months. Harvesting is done either by lifting the poles or by divers who are able to select market size mussels leaving the undersized mussels on the poles for further growth.

In the Philippines, yield may reach 15 kg/4–m pole. An area of 1 ha with 20,000 poles is capable of producing 50 t of mussel annually. This is a simple direct method of mussel culture and in the tropics should be the first one considered. Depth of pole plantings and season are critical factors that require investigation.

STOCKING METHOD

The advent of relatively inexpensive synthetic mesh tubing has made possible a new method of mussel culture, as well as permitting variations in older types such as the tubes in the French bouchot system. There are several types of tubing such as the brand name Netlon with an open flexible mesh in four sizes (Italian manufacture), a Norwegian closely knit tube, and a further modification manufactured in Canada. There is also the standard fixed mesh. The brand name Vexar is one type.

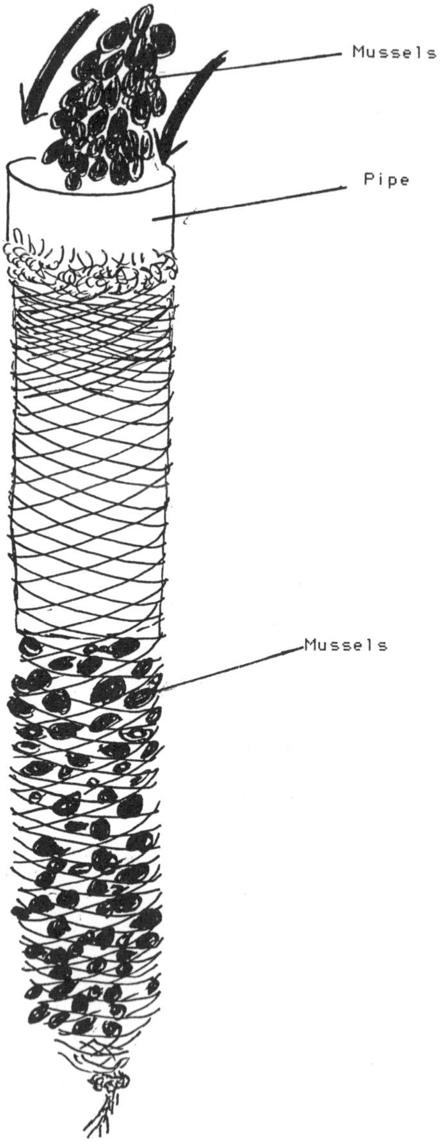

Fig. 89. Filling mesh tubing with mussel spat.

The tubing comes in endless rolls that may be cut to a desired length. A firm pipe of the appropriate diameter is inserted in the mesh tubing and filled with mussels, after which the pipe is withdrawn (Fig. 89). The tubes are suspended either from rafts or long lines in the same manner as oyster strings. The mussels eventually find their way to the exterior of the tubing.

Before placing the mussels in the tubes, it is of advantage to allow them to clump together, if this has not already occurred. This may be done by placing them in a container with water for a few hours. Filling the tubes is accomplished better with clumped mussels. The system is particularly applicable where spatfalls are of low intensity or where seed has to be transported to growing sites, as from beach sets, and for small mussels, such as those too small for marketing after harvest. Various sizes of the Netlon type permits resocking as the mussels grow, but this is not done with Norwegian or Canadian type tubing. In these tubes the usual size of seed mussels is 10 to 15 mm. One metre of such tubing can produce about 10 kg of 50–mm mussels.

SYNTHETIC MESH CULTURE

Semirigid netting (e.g., 12.7– mm mesh Vexar) cut into strips 10 cm wide and of appropriate lengths (usually 2 m) are suspended from rafts or long lines for mussel seed collection. A loop is made and a small weight such as a rock is tied to the bottom of the loop to counteract the buoyancy of the netting. After settlement of the spat and some growth, the rock weight is removed and the loop released to allow the strip to hang at its full length of 2 m.

The density of settlement may be controlled to a certain extent by the size of the mesh; the smaller size will collect more, and larger mesh less. Such collectors can have a life of 10 years or more if handled carefully. Discarded fishing net, though not as durable, will serve the same purpose.

After seed collection and a size of about 15 mm reached, seed may be re-set into mesh tubing. If the mussels are allowed to grow to maturity on the collector, a mesh size of 25 mm is required. Harvesting mussels from Vexar sheets may be done with a hose at a water pressure of 3 - 4 kg/cm^2.

MUSSEL CONDITION FACTOR

The condition factor of mussels may be determined exactly as for oysters. However, to extract the meat more expeditiously, it is advantageous to steam them open. Either the actual weight of the steamed meats or the meat dried to a constant weight may be used for the calculation. Four duplicate samples of about 25 mussels each is a suitable number for the condition factor determinations of mussels.

Mussels in the ripe spawning condition attain the highest condition factor and the lowest is reached after spawning. There is only a brief resting period after spawning before proliferation of new gonadal material.

MUSSEL CULTURE PROBLEMS

As with oysters in suspended culture, fouling may be a significant problem which may be dealt with in the same way. In the tropics, rapid mussel growth, usually half a year to market size, may make it possible to culture around some of the fouling problems.

Another difficulty with mussels is the byssus which is, of course, inedible. With fresh market mussels, customers are usually responsible for byssal removal, but processing, such as canning, requires a byssal–free product. Machines are available for byssal removal.

Still another problem is that of the shelf life of suspended culture mussels, as they have a tendency to gape soon after removal from the water. The solution is either adequate refrigeration or acclimation to time out of the water by gradually increasing air exposure. This may be done by intertidal storage.

Predators are mainly fish, seastars and birds. Seastar predation depends on the timing of culture operations and how they coincide with seastar breeding, especially with suspended culture. If the timing is right, seastars may not be able to attain a size large enough to do damage to the larger mussels. There may be an attempt to reduce the population of fish by fishing operations or by protection of the culture site with a protective net, which will also prevent access by birds.

HARVESTING

Bottom cultured mussels are harvested by raking at low tide or dredging at high tide. In lifting strings from suspended culture, care must be taken to prevent mussel clusters from dropping off the strings or stockings. Next the mussels should be washed. This can be done with a high pressure hose or a rotating drum with a sprayer. Grading is done according to size and the amount of fouling, particularly barnacles.

As with oysters, mussels may be stored either wet or dry and in a similar manner either in sink floats, intertidally on the bottom, or in submerged open mesh sacks. Dry storage requires refrigeration, either with ice or in a refrigerated room with a temperature between 2 and 5° C. At this temperature, mussels can be kept in good condition for 2 or more weeks.

For shipping, mussels may be conditioned by holding them intertidally for a time that will depend on the tidal range and tidal cycle. They are shipped in the dry state, preferably in refrigerated transport.

PROCESSING

Most mussels are sold fresh to the consumer. However, they may be processed in several ways. Steaming will release the meats from the shells without individual shucking. They can be canned in a light brine or in mussel nectar which is the juice collected in the steaming process. Mussel meats can also be smoked and canned in a light oil. They are also pickled in a light vinegar with a variety of spices. In Asia mussels are steamed and the meat is dried.

CHAPTER 8

CLAM CULTURE

Since there are so many species of edible clams, it is impossible to cover the culture of all of them. Therefore, the culture of only three species, each quite different from the others, will be examined. The little neck clam (*Tapes phillipinarum*) (Veneridae) is widely fished but with limited culture in the Indo Pacific and west coast of North America. Similar species occur and are cultured in Europe, particularly in Spain and Portugal. (*Mercenaria mercenaria*, the quohog [also a venerid] is fished and cultured on the East coast of North America with similar culture practices.) The ark shell (cockle or blood clam), *Anadara*, is widely used in tropical countries, particularly in Southeast Asia, where a semi-culture of *Anadara granosa* is most productive. The third species with potential in the tropics is the giant clam, *Tridacna*, whose culture is in the developmental stage.

Hatchery operations for all of these species closely resemble that of the oyster, so the common features will not be repeated. The habitats of the three species are quite distinct and represent examples of the more dominant habitats of edible clams. While some forms of the semi-culture of clams have been carried on for many years, clam culture in the fullest sense is relatively recent and the systems are still in the formative stages, as compared, for instance, to the culture of oysters.

SPAT COLLECTION

Oyster cultch may be used to collect clam spat, but since the byssal attachment is of short duration and young clams have the ability to move off the cultch, frequent examination is necessary. The most typical approach to collecting clam spat is to sieve the surface layers of clam beds. This requires a series of sieves with different mesh sizes whose gradation depends on the particle size of the sediment. The object is to screen out very fine particles like mud and sand grains and larger pebbles. Even when these are eliminated, stereoscopic microscope examination is necessary to pick up the smallest specimens of clams.

A difficulty not usually encountered with oyster spat, where the species are few, is the identification of clam spat. Various species of clam spat may be quite similar when very small, and it may be necessary to grow those from the collection to a size when adult characteristics become apparent and established. This can be done in fine meshed trays. If the larva of the species of concern is known, the size and shape of the prodissoconch seen in the juveniles may assist in identification. In any case, this should be recorded for each specimen in the collection and may be a means of initial separation of species.

GROWTH STUDIES

Growth studies of young clams can be done in trays, but most clam species do not survive well in trays, out of their natural habitat, as they require the pressure of the substrate in which they live. Compressing the two valves with elastic bands may assist in prolonging survival in exposed clams. However, growth studies should be carried out in an environment as natural as possible. Depending on species, younger stages may be held in trays of sand or gravel, but the natural substrate is the ideal.

CLAM CULTURE

Clam culture has not attained the status of oyster culture, partly because clams generally do not lend themselves very well to manipulation. Another factor is competition with wild clam stocks which may be represented as follows:

Clam culture costs = seed collection + planting + harvesting
Wild clam costs = harvesting

It can be readily seen that, where wild stocks occur in sufficient abundance, the additional costs of seed collection and planting may make culture non-competitive. Also, until recent years the value of clams has been relatively low, thus affecting the economics of clam culture. Notable exceptions are the culture of *Anadara* in Malaysia and Korea, *Tapes* in Puget Sound, Washington, USA and *Mercenaria* on the east coast of the USA.

Until the recent advent of shellfish hatcheries, clam farming had been termed a semi-culture, simply involving harvesting of naturally set wild clams of small size and transplanting (at the same time, concentrating) them to another area, presumably better suited to growth or for easier harvesting. More efficient harvesting is partly a result of high density planting. This system of transplant culture has long been used in China for the razor clam *Sinonvacula*, the so–called cockle, *Anadara* and the clam, *Tapes*. Another species, *Tapes philippinarum*, as well as native *Tapes* species, are similarly semi-cultured in Spain and Portugal and formerly in Japan. More recently (the last 40 years), the blood cockle (*Anadara granosa*) has been cultured in this way in West Malaysia, Indonesia and Thailand. In Japan, a number of species of bivalves have been cultured on the transplant system but none has reached the sophistication of oyster culture, except for recent developments with scallops.

Hatchery production of clam spat has been developed to a point where quantities are large enough for commercial culture operations to develop in a number of areas. Among the species being used are *Mercenaria mercenaria* on the east coast of the USA and *Tapes philippinarum* in China and the west coast of the USA. Culture of some species of *Tridacna* from hatchery spat is in preliminary stages in the Southwest Pacific.

As with oysters, the use of hatcheries is predicated on low abundance and availability of seed in nature. Clam seed of most species, as indicated, is more difficult than oyster seed to collect in abundance and at low cost.

LITTLENECK CLAMS

This littleneck clams include several genera of relatively small clams in the family Veneridae. Among the main genera are *Paphia, Venerupis* and *Tapes* which occur in both tropical and temperate waters. The usual size at harvest is in the range of 3 to 4 cm (less in Japan) or 40 – 60/kg. They are steamed and eaten directly out of the shell or canned, although they may be eaten raw on the halfshell.

Other than in areas where labour is relatively cheap, collection of wild seed for replanting is virtually out of the question. It may be possible to increase seeding of both cultured and uncontrolled beds by erecting barriers such as brush fences at right angles to the current flow. This provides a collecting surface for byssal attachment of the larvae. When the spat lose the byssal attachment they drop to the ground and will bury themselves.

Unless economic conditions in the area permit, most little neck culture in the future will be based on hatchery seed. However, the culture procedure for both types of seed is identical. Most littleneck ground should be muddy, or sandy, gravel with pebbles in the 10 mm range. The main criterion for bottom sediments is stability. Sand alone seldom supplies this. The ground should be reasonably well protected from wave action and moderately free of predators such as seastars, moon snails and tritonalian type borers. Some species such as *Tapes philippinarum* live quite high in the intertidal zone, while others such as *Paphia* spp. live lower down or even subtidally. Wading depth subtidal clam culture is possible.

The tidal level at which clams occur naturally is used as the starting point for determining the tidal level for culture, and experimentation will determine the extent to which this may be broadened either up or down the beach, or both. Beaches with a steep slope (10 to 1) are more subject to erosion and loss of young clams than a gentle slope (20 to 1).

Presence of clams at the site usually indicates temperature and salinity conditions are satisfactory. If no clams are present in the proposed site, it will be necessary to make measurements of temperature and salinity, bearing in mind that the source of fresh water may fluctuate throughout the year. The data will determine if the range is within the limits known for the species. Most clams can withstand short periods of low salinity. The littleneck clam, *Tapes philippinarm*, has an optimum salinity range for growth between 20 and 30 ppt, although it can tolerate salinities as low as 10 to 12 ppt for periods of a month.

It is advisable to divide the area into plots, particularly if a considerable tidal range or substrate variation exists. This will make for better productivity records, permits some experimentation with planting densities and provides a systematic approach to planting and harvesting. Plots not more than 10 m x 10 m are a suitable size. Plots may be prepared for planting by raking, lightly loosening the surface layer of the bed especially if it is a compacted one. This is often done just prior to expected settlement in growing areas not already under cultivation.

A recent development in clam culture is the use of netting as a form of surface protection for young clams. The beds are covered with netting of approximately 12 mm mesh size. A mesh of smaller size tends to become clogged by algae and silt and larger mesh, above 25 mm, gives insufficient protection. The most suitable material for long life is synthetic such as the Vexar commonly used in aquaculture for many purposes. A particularly suitable type is one known as "carcover." However, used fishing nets of appropriate mesh size would be suitable.

In addition to protecting planted clams, it has been found that cover may increase the density of natural sets not only by the protective nature of netting, but by providing a site for byssal attachment by larvae and prevention of erosion of the beach surface.

The netting is prepared by sewing pieces together to form an area large enough to cover a plot (one reason for small plots). Before laying the net, a shallow trench is dug completely around the plot perimeter. The net should be large enough for edges to fall into the trench. Stakes are driven through the stretched netting at 3– or 4– m intervals to hold it in place, after which the trench is filled in with soil. Whether the net is placed before or after seeding depends on the size of the seed clams. If larger than the mesh size, they must be planted before laying the net. If smaller, planting can take place after the netting has been put in place.

The clams should be planted on a rising tide, at a time when there is minimal wave action, with care taken to prevent the seed being exposed to the sun for a long period. Depending on the beach slope, it may be possible to spread seed when covered with a few centimetres of water. In a sub-tidal area (wading depth), the netting can be held in position by stakes alone, driven in at smaller intervals with the outline of the plots being shown by long stakes or buoys. All of these operations are timed according to the tidal regime.

SEED CLAMS

It is important to take good care of seed clams. Hatchery seed should be placed in water as soon as they arrive at the planting area. They should be held either in fairly large containers of fresh sea water kept at the temperature of the water where they will be planted, with the water either aerated or changed several times a day. Local wild seed may be planted immediately, kept in containers as above or in mesh bags in the sea. Hard clams (little neck clams) are usually in the 3– to 4– mm size range when received from the hatchery. It is difficult to obtain wild seed less than 10 mm in length.

Numbers of clams being seeded should be known as accurately as possible and this may be determined by counting the number in a known volume or given weight and relating these to the total volume or weight of seed. A minimum yield at harvest of 100 clams 30 mm in length/m^2 might be expected and experiments have shown that at least 500 small (3– 4 mm) seed/m^2 are needed. If larger seed (10 mm) is used, proportionately less is required. Initial experimental plantings of different densities in different plots should be made.

An experimental design suitable for such experimentations that permits standard methods of statistical analysis is a 5 x 5 Latin square. Twenty five plots of equal size are arranged in a square on the beach. The size will depend partly on the beach area. Plots may be as small as 1 m^2, but preferably larger, up to 10 m^2. There will be 5 treatments or planting densities, for example, of 200, 400, 600, 800 and 1000 seed/m^2. Each one of these treatments occurs only once in each row and each column. There can be many such arrangements and one should be chosen at random. One such arrangement is shown in Fig. 90.

To compare different beaches, a similar square could be placed on each and statistical comparisons may be made. This arrangement of only one treatment in each row and each column will show differences in fertility up and down and across the beach. Data on both productivity and growth will be obtained. The data may be analyzed either with the Chi square method or by analysis of variance to determine the significance of differences between treatments. If necessary, a statistician may be consulted on design and analysis of results of such an experiment.

A	B	C	D	E
C	D	A	E	B
E	A	B	C	D
B	E	D	A	C
D	C	E	B	A

Fig. 90. A 5 x 5 Latin square.

Problems of clam culture, apart from seed procurement, occur with movement of clams from the planting area, predation, propellor wash, pollution and poaching. Movement of clams from the plot is not affected by the clams' ability to move, which is essentially the "digging in" activity, but from wave or current action. This is where net protection is of value. Small clams are more subject to dispersal than large ones. Another source of movement results from the propellor wash of motorized vessels crossing over the plots at high tide.

PREDATION

Predation is quite similar to that for oysters with seastars and crabs a common problem. In addition to the triton type gastropod drills, there is also the naticid clam drill. These live and move below the surface and drill in the same way as the tritons. The drill hole of the naticids may be identified by the fact it is counter-sunk (Fig. 52) or bevelled, while that of the tritons is not. The difficulty with naticid drills is that they cannot be seen except during the egg laying season when they deposit sand collar (Fig. 52) type egg capsules. The snails usually come to, or close to, the surface at this time. Removal of the sand collars will limit population increases.

Fish, particularly bottom feeders such as flat fish like soles and rays, prey on small clams and the only control possible is a regular fishing program or fences. Birds are an important factor in clam predation and gulls along with ducks are mainly responsible. Deterrents such as the line of flags used in rice paddies may assist.

FOULING

Fouling may also occur, causing the openings in the protective net to be covered. Barnacles and mussels, along with several species of algae such as *Entermorpha*, may become problems. The latter may form dense mats. If the area is not too large, sweeping or raking may reduce the worst effects of such fouling.

HARVESTING

Time of harvesting depends on growth rate and market demand of various sizes. Harvesting method varies with the species, particularly the depth at which it lives. Hand harvesting utilizes implements shown in Fig. 91. Mechanical harvesters use water jets to move the clams to the surface where they are picked up. This is the best way to dig clams since sediments are returned in their proper layers. There is also little or no shell breakage and smaller clams are not buried. In hand harvesting, care should be taken not to mound the soil, rather it should be dug as though digging a garden for planting vegetables. The smaller clams may be buried too deeply to reach the surface causing some mortality. Considerable shell damage and mortality is possible.

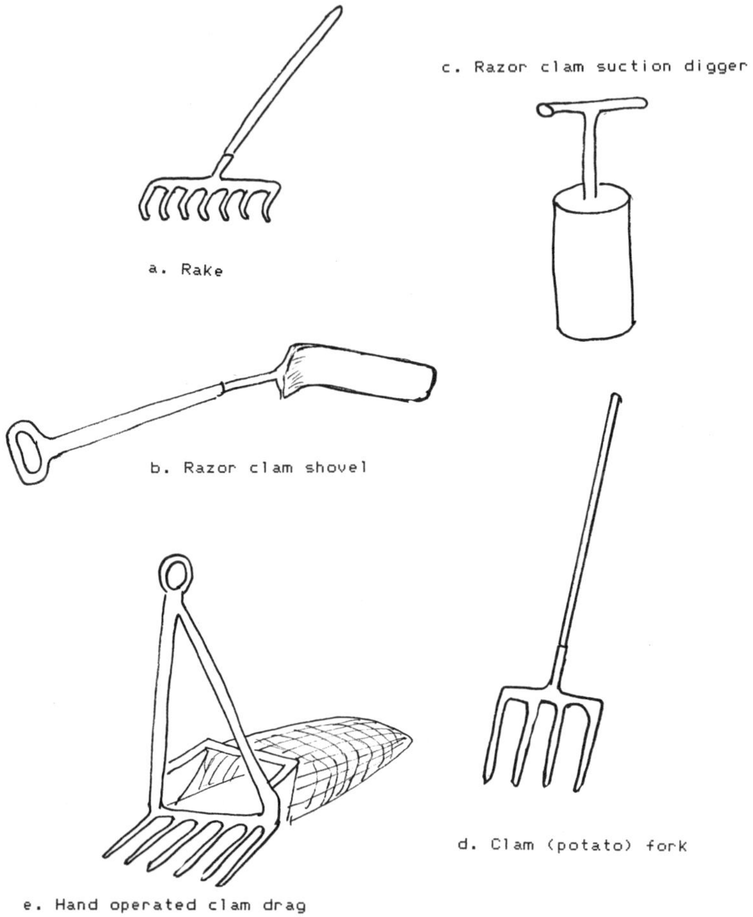

Fig. 91. Clam–digging instruments.

After harvesting, clams may be stored in sink floats as described previously for oysters. This gives the clams an opportunity to be cleansed of sand and silt. They may be stored in layers to a depth of about five times the extent to which they can extend the siphons. If stored for a lengthy period, dead and dying clams should be removed at frequent intervals. They are better stored in such a facility rather than in sacks under water unless they are to be held for only a day or so. If sacks are used, they should have open mesh such as onion sacks rather than closed mesh jute.

PRODUCTIVITY

Productivity will vary with the species, the time allowed for growth, growth rate, the planting density and the survival rate. For littleneck clams a production of 4 – 6 kg/m^2 is possible. For large species, such as *Saxidomus*, 30 kg/m^2 have been taken from the wild ground.

The economics of clam culture will depend on the area, labour costs and value of the species. An approximate breakdown of proportionate culture costs is as follows, with either natural or hatchery seed:

Seed	45%
Netting	7%
Stakes, etc.	2%
Labour	7%
Seeding	7%
Maintenance	2%
Harvesting	30%

By comparison, in bottom oyster culture seed costs may be 28% of the total and harvesting 42%.

THE COCKLE (*ANADARA*)

One of the most productive clams in the world is the so-called cockle of the genus *Anadara*. There are many species, the most prolific of which is *Anadara granosa* which occurs in Southeast Asia (Indo-Pacific). Nearly all ark shells (*Arcidae*) are tropical but the Noah's ark shell (*Arca tetragona*) is found in Great Britain. A common, much utilized species in West Africa is *Anadara* (*Senilia*) *senilis*. There are reputed to be 15 species on the Pacific coast of Colombia in South America. Some are quite large with heavy shells like *A. senilis* (14 cm) and a few have byssal attachments.

The shell of *Anadara* spp. cannot be mistaken because of the deep cockle-like ribs and taxodont hinges on which there are numerous chevron shaped teeth on both sides of the umbone (Fig. 92). The umbones are more

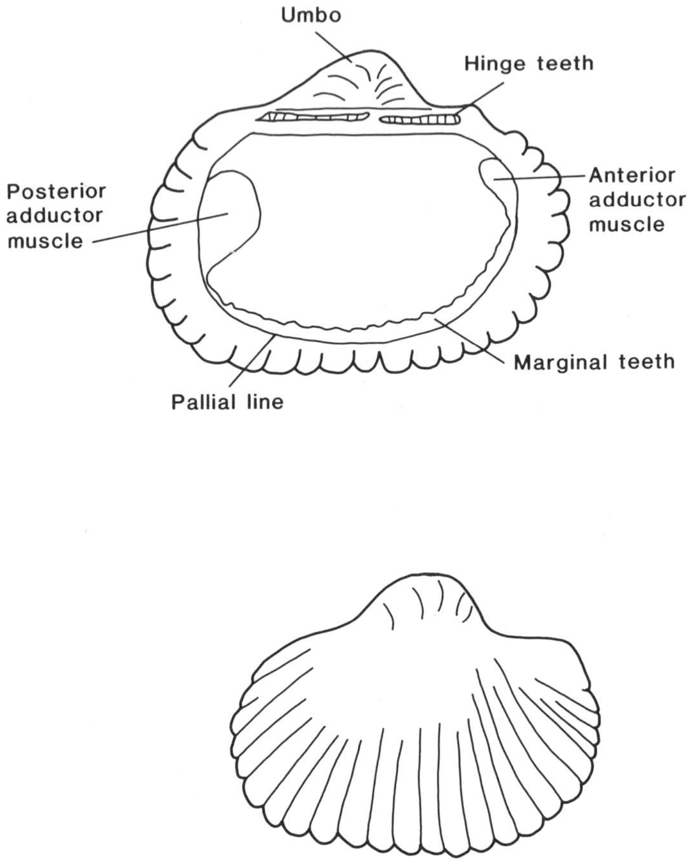

Fig. 92. Shell of Anadara granosa.

or less centrally placed. Since there are no siphons, there is no pallial sinus, and in *Anadara granosa* the posterior muscle and scar are larger than the anterior.

Because of the red-brown colour of the flesh *Anadara* are sometimes called blood clams or cockles. The colouring is caused by haemoglobin in the blood. Being without siphons, the inhalant flow enters the anterior ventral shell area and the exhalant flow goes out just below the posterior adductor muscle. The foot is well developed, enabling the cockle to travel easily through the fine silt of the habitat.

Biology

These species occur in estuarine mud flats bordered by mangrove forests. The habitat of the *A. granosa* type cockle is very soft, almost liquid mud, up to 75 cm deep, usually rich in organic material, at lower tidal levels and into the subtidal. Such areas are typically found in mangrove estuaries. Being an estuarine habitat, salinity fluctuates greatly even during the daily tidal cycle and especially during the rainy season. The normal habitat in West Malaysia has a salinity between 18 and 30 ppt and temperature in the 25 to 30° C range. Waters overlying the cockle beds are nearly always turbid, both from the silty bottom and from phytoplankton production. The tidal range is 2 m.

The centre of concentration of *A. granosa* is in Malacca Straits between West Malaysia and Indonesia. Up to 65,000 t annually are said to be produced in this area (Malaysia and Indonesia) from wild stocks and from a transplant culture. The species occurs as far north as southern Japan and as far west as India.

The gonad develops when cockles reach about 18 mm in length, and first spawning seems to occur at the 25 mm mark. There is some spawning in the Malacca Straits throughout the year, with a peak in the latter part of the year.

The larva is quite distinctive and similar to most other Arcidae larvae. The larval period appears to be a long one of at least 3 to 4 weeks. *Anadara* larvae are much longer than high, more so than most lamellibranch larvae with a length/height ratio of 1.13 and a setting size of about 230 µm. The umbones are long and broad and the anterior end is slightly more pointed than the posterior, though both are rounded. The hinge teeth are small and numerous with the ligament toward the narrow end. Living larvae of *A. granosa* are yellow in colour.

The spatfall may be as dense as 4500/m^2 and this occurs at about the low water neap tidal level. The average density is considerably less.

Growth

Growth of *Anadara granosa* is rapid, and it reaches a length of about 5 mm 1 month after spatting. Seed is sown when 5 to 10 mm in length and, at low density, may reach market size of 32 mm in 8 to 10 months, but increased density lengthens this time.

Culture

Seed is harvested at low tide by drawing a hand held basket through the surface layer of mud while kneeling with one knee on a board for support in the soft mud and pushing with the other leg. A box is held on the board for the collection. One person is able to collect 5 litres or 5 - 8,000 cockle seed/ hr. Seed collected by fishermen are sold to lease holders which are sometimes organized as cooperatives.

Growing beds are generally farther down the estuary where there is little or no spatfall. If the transfer is local, the seed transferred from the fisherman's sampan to the larger vessel from which the seed is shovelled overboard onto the beds marked by buoys or stakes. Seed may be shipped long distances in tins. Sometimes it is held in an intermediate nursery area before final planting. Planting density is approximately 1500 - 2000 spat/m^2 resulting in a final production of about 800 clams/m^2. Market size (legal) is 32 cm with a weight of 10 g giving 8 kg of cockles/m^2. The cockles are harvested by drawing through the mud a hand held basket (larger mesh than the seed basket) on the end of a pole (Fig. 61). This is usually done from a boat at high tide. An alternative is to hold the basket and pole firmly fixed to the vessel and the movement of the powered boat draws the basket through the mud. Catch per man-hour may be 150 kg. The cockles are shipped in plastic sacks containing 65 – 70 kg. Production in West Malaysia along the Straits of Malacca varies around 65,000 t annually, most for local use but some is exported.

TRIDACNIDAE

The culture of *Tridacna* species is just developing in Micronesia. One attraction of this group is its ability to grow well in waters of low productivity. The culture methods have not been developed very far at this time. Current literature should be consulted to determine the latest developments.

Hatchery Production of Seed

The only source of tridacnid seed clams at this point is from hatcheries. For spawning in mass culture clams are maintained in large tanks (8 x 2 x 1 m). All normal methods of stimulation of bivalve spawning are used but the most effective appears to be mascerated gonads, although freeze dried gonad also appears effective. However, natural spawning is preferable. In nature, free gametes stimulate others to spawn. Males usually spawn first and the tank is drained immediately to flush out excess sperm. The tank is refilled to allow release of female gametes. The density of ova is adjusted by transfer to another tank to attain a level of about 1000 ova/litre. In about 1 week, the veligers have spatted with byssal attachment on the sides of the tank which has held unfiltered sea water with no additional food. In small scale laboratory culture, the standard larval food is *Isochrysis galbana* and the larvae appear to be able to ingest and absorb free living zooxanthellae.

The larvae of tridacnids are nondescript lacking any distinctive characteristics and with only slight indications of umbones. The size at settlement varies between 175 and 200 µm in length in a larval period of 9 to 12 days at a temperature of 30° C. The spat are thigmotropic or even cryptic, seeking out corners or clefts.

Spat are allowed to remain attached to the tank walls for about 90 days when they are 5 mm in length. They are transferred to trays (Nestier or similar type) at a density of $1000/m^2$ in troughs with a slow flow-through current. The trays also contain cultch pieces of firm materials such as cement or tile chips about 1 cm in diameter for byssal attachment of the spat. Thinning is done as the clams grow.

Spat of most tridacnid species grow up to 2 cm in about 5 months. This nursery stage is the critical factor in tridacnid culture for, like hard shelled clams (*Tapes*), they cannot be placed in the wild without protection until they reach a size of about 20 cm. This means some protection in the form of trays in the open or inshore holding installations, all of which are costly. The economic solution to this problem will probably determine the feasability of large scale *Tridacna* culture.

Growth

Along with the planktonic sources of food, as with most clams, the tridacnids have the additional source of nutrients from the zooxanthellae in the siphon-mantle complex. The large size attained by tridacnids is attributed to this auxiliary source of food. Zooanthellae are acquired as an auxiliary food source soon after metamorphosis. Good growth data are not available for tridacnids, but a rate of 5 cm/year (20 cm length at age 4) for *T. deresa* is comparable or better than the rate for *Crassostrea gigas*, known as a rapid grower. At 12 cm, *T. deresa* can have a wet meat weight of 45 g while an oyster (*C. gigas*) of equal length and age may have a meat weight of 30 g. This growth rate can be obtained in nutrient poor waters where other bivalves would grow very poorly. It is calculated that 3–year–old laboratory cultured *T. deresa* grown at a density of 50/m^2 could produce 16 t of edible meat/ha with about 3 t of it as adductor muscle, the most valuable part, along with 70 t of shell. It has been demonstrated that 20–mm tridacnid spat may be effectively shipped long distances.

Predators

The tridacnids have the usual quota of bivalve predators. Drills are able to attack through the byssal aperture as well as through the shells of smaller specimens. Crabs and octopus are also common predators, as well as fish, on smaller specimens.

CHAPTER 9

SCALLOPS

Scallops, also called clams in Great Britain, or Coquille St. Jacques in Spain and France, belong to the family Pectinidae, in which there are several hundred species. Many are quite small, about 10 mm in diameter, while the largest measure up to 30 cm in length. The number of commercial scallops number only a dozen or so. The main commercial genera are *Pecten, Placopecten, Patinopecten, Argopecten* and *Chlamys.*

Scallops are cultured in a few places. The Japanese have been culturing *Patinopecten* for many years while limited efforts have been made in Europe with *Chlamys* and in North America with *Argopecten.* Since the large El Nino of 1982 and 1983, small–scale scallop culture has developed in Peru and Chile with *Argopecten purpuratus.*

CONDITION

It may be recalled that condition factor in oysters is measured by comparing meat content to internal shell volume. This method of measuring condition factor cannot be applied to scallops because of the open byssal aperture, allowing air to enter the shell cavity and altering the underwater weight. However, an approximation of the internal shell volume is the difference between the whole weight and the shell weight in air.

As in other bivalve species, the meat content increases with gonadal development and falls drastically after spawning. This is followed by a resting period during which food reserves accumulate, which in most bivalves are concentrated in Leydig tissue but in scallops they accumulate to a large extent in the adductor muscle.

Since the adductor muscle is important commercially, some importance is attached to its size and condition, and several indices have been developed for it.

1. $$\frac{\text{Muscle diameter} \times 100}{\text{Shell height}}$$

2. $$\frac{\text{Muscle volume} \times 100}{(\text{Shell height}) \times N}$$

where N = the slope of the log (volume)/log(height) relationship.

A measure of total condition may be determined by:

1. $$\frac{\text{Weight of soft body}}{(\text{Shell length})^2 \times \text{Shell breadth}}$$

2. $$\frac{\text{Wet meat weight} \times 100}{\text{Total wet weight}}$$

In addition, the gonadal index may be used as a measure of condition as well as of gonadal maturity, for this is the organ that undergoes the greatest annual variation in weight. The gonadal index is:

$$\frac{\text{Gonad weight} \times 100}{\text{Total weight of soft body}}$$

Unlike most bivalves the scallop gonad may be removed almost intact from the rest of the body.

SEED COLLECTION

Since virtually all scallops occur subtidally, some to the depth of 200 m, collection of seed presents problems somewhat different from oysters which are primarily intertidal or shallow water species. The depth at which scallops live implies larval distribution to that depth so some knowledge of the vertical distribution of larvae and spat are required. This is done in the same way as for oysters but with an extended vertical sampling program. While on an experimental basis oyster cultch can be used, the commercial scallop collectors take a different form. The collector type found most efficacious is porous bags such as onion sacks filled with monofilament synthetics. Used nylon fishing nets are suitable. Such collectors may have a life of several years.

The mesh of the container bag should be less than 5 mm so any spat falling from the collector threads will be caught in the bottom of the container. Onion bags are about 100 x 50 cm and in Japan these have collected up to 50,000 spat emanating from a larval concentration of 1000/m^3. A satisfactory commercial set is in the 500 – 1000 spat range from larval concentrations of 100 – 200/m^3.

The necessity for spatfall forecasting will depend on the local situation: whether spawning is within a short period or is prolonged, the size of the scallop population, the type of larval distribution, both vertically and horizontally, and the degree of fouling to be expected on the collectors. Several years of basic data collection are required before the need and possibility of forecasting becomes clear.

Suspension of collectors is by raft or long line and since, in some cases, spat settlement may be well below the surface, a method of submerging the collectors at a specific level is required. This is done by sinking a long line, still suspended by floats, below the surface to the required depth (Fig. 93). This system requires lifting the anchors to work with the scallops. Another system also used to sink a long line below ice is to install a second set of anchors (Fig. 93), one set for a surface operation and another set for submergence. The secondary anchors must be lifted but the other anchors maintain the position of the long line, not possible in the first system.

Collectors are attached at intervals on suspension lines, each of which must have a weight at the end, at the depth span shown by experiments to produce optimum settlement. The size of the equipment, such as rope diameters and anchors, are as for oyster long lines but the seed suspension lines can be of a smaller size, possibly 3 to 5 mm.

NURSERY

Spat are allowed to remain on the collectors until they attain a size of 5

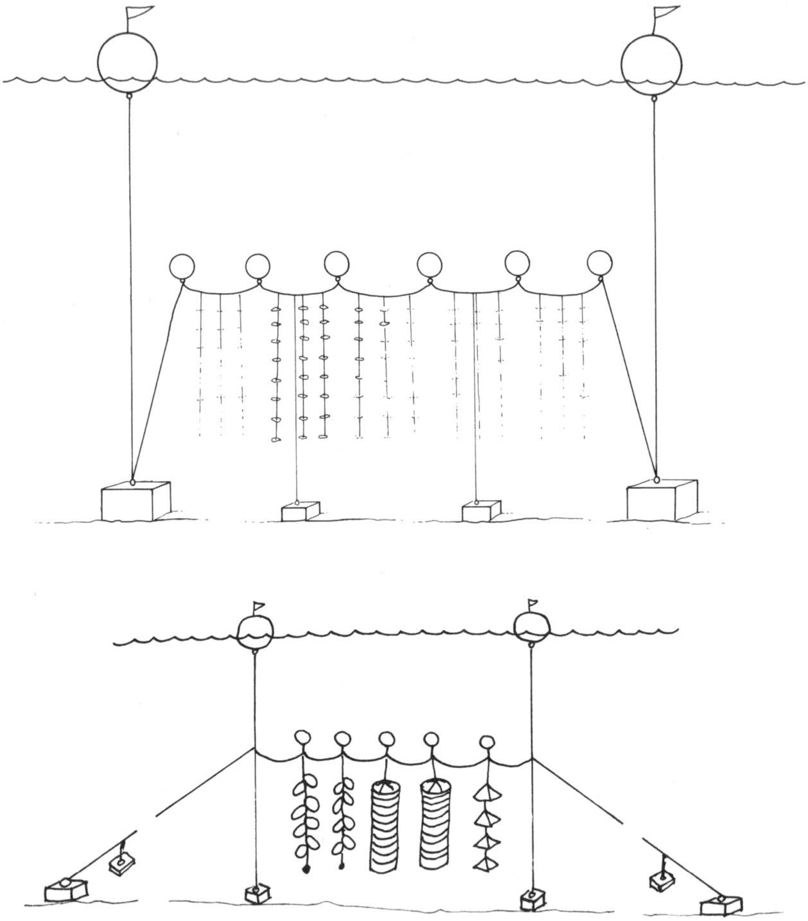

Fig. 93. Sunken long line with main anchors on long anchor lines and supplementary anchors which can be hauled to the surface to work on the long line.

to 10 mm when they are removed to a nursery or intermediate culture situation. This usually takes the form of a tray. The trays described for oyster culture serve the purpose. Manufactured netting trays called pearl nets or lantern nets (Fig. 86), which were originally made for scallop culture, are obtainable with meshes of various sizes. Most pearl nets are 30 – 40 cm in diameter or square with meshes in intervals between 4 and 12 mm. The stocking density of trays or nets is a function of the size of the spat, but 10– mm can be placed at 1 spat/3 cm^2 and 25–mm spat at 1/25 cm^2 of tray area. Scallops do not tolerate as high a stocking density as other bivalves. As spat grow they should be transferred to nets or trays of larger mesh size at reduced densities. This permits fouling clearance at the same time.

ADULT CULTURE

When spat reach a size of approximately 3 cm, they are ready for the next culture stage sometimes called the "grow out" process. Among the various systems are bottom (sowing) culture and suspended systems such as pearl nets, lantern nets, pocket nets and ear culture.

Bottom Culture

The factors influencing bottom culture are sea bed consistency, depth, currents and predators. Scallops are cultured on the bottom as deep as 30 m but this requires vessels able to pull harvesting dredges over the bottom. However, many tropical species grow in relatively shallow water within the reach of divers or oyster tongs.

The bottom consistency of most natural scallop beds is one of firm sand, gravel or a mixture of the two. A limited amount of mud is tolerated. Some species of *Chlamys* inhabit rocky bottoms. There appears to be a preference for some current flow but with not enough speed to unduly disturb the bottom sediments.

Predators include the usual forms associated with clams and oysters and include seastars, drills, crabs and fish. Control measures are also similar. Seastars are a major predator and the beds may have to be dredged or treated with lime to reduce the population before planting seed.

For large scallops in deep water, such as *Placopecten* and *Patinopecten*, a final concentration of about 5 scallops/m^2 is considered suitable from a consideration of growth rate and of harvesting efficiency. Planting seed at least 3 cm in diameter should be done at a density of 15/m². A mortality rate of up to 50% or more may be expected, and harvesting by dredges is not 100% efficient. In temperate waters these species require at least three years on the bottom to reach market size. The harvest of deep water scallops is carried out by dredges. There are many designs of dredges, in part depending on the type of bottom from which the scallops are being harvested. Most are of a chain link variety, but in some cases shrimp trawl nets have been modified for the purpose.

Tropical species, generally smaller in size (between 5 and 10 cm in maximum length), require a little over a year to reach market size. Culture of smaller species of scallops is just beginning. One species, *Argopecten purpuratus*, which occurs in relatively shallow water of 2 to 10 m is being cultured on an artisanal basis in Peru. Seed, about 3 cm in length, are collected from natural beds and transplanted to leases about 1 ha in extent in a growing area with a depth of 3 to 5 m. Leases are marked off with surface buoys. They are enclosed with 1 to 2 m high gill nets with the lead line (bottom) additionally weighted with small rocks to hold the net on the bottom; the nets are held erect by normal cork lines (Fig. 94). Stocking density is high, up to 100/m². These scallops grow at a rate of 1 cm/month and are harvested by diving when they are 7 to 8 cm in length. In such shallow water they could be harvested with tongs but are usually gathered by divers.

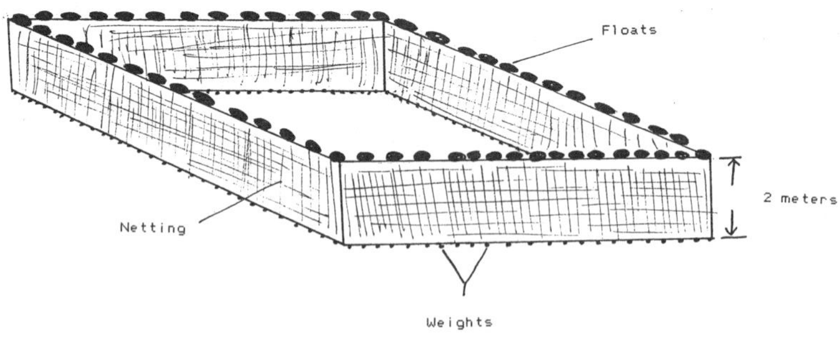

Fig. 94. Peruvian type of net enclosure for scallop culture.

Suspended Culture

Pearl nets, lantern nets or other trays of various types are used for adult culture as well as for nursery purposes. Pearl nets were specifically designed for seed, while the lantern net was for final culture. Suspended culture has an advantage of a more rapid growth rate than bottom culture. It provides a measure of control over the population, allowing some manipulation such as culture depth and density. It can be used in areas where the consistency of the sea floor does not allow bottom culture, and it can be operated above bottom culture. Also predation is reduced to a minimum. However, space on the nets is costly and, as explained previously for tray culture of oysters, there must be a compromise between growth rate and optimum use of space.

A rule of thumb requires that scallops do not take up more than one–third of the tray area for optimum growth. Also with scallops, overstocking can cause malformed shells, although this does not necessarily affect the meat. There may be mortality from two scallops locking (biting) together, with the ventral margin of the valve of one scallop inside the two valves of the other. Unless the initial concentration of scallops in a net is low, thinning is required and this is labour costly. With the cost of nets or trays and the long line or raft, suspended culture is costly. However, hanging culture with a 2 year culture period may be more profitable than bottom culture with 3 years for grow out.

So far, lantern nets are the main culture apparatus for suspended culture, other than the long line or raft itself. These are 7 to 10 wire hoops 15 to 20 cm apart enclosed by monofilament of 12 to 20 mm mesh size. Each hoop of 50 cm in diameter is meshed and forms a platform for holding the scallops. A net with 10 hoops can make a column 2 m long, and the whole apparatus can collapse like an accordian for storage. Several lantern nets may be hung one below the other if there is sufficient water depth. The number each platform holds depends on the scallops' size. With the one-third area rule, 1980 cm^2 of the platform should hold no more than 10 scallops of 10 cm diameter, or each lantern net of 10 platforms has 100 scallops. Thus, significant production requires a considerable quantity of nets. Pearl nets are constructed in a pyramid shape and may be hung one below the other.

Another alternative is the pocket net, 75 to 100 cm high, and 45 to 60 cm wide, made of 3–cm synthetic mesh on a 5–mm wire frame. On this frame are sewn rows of pockets to hold the scallops (Fig. 95). The frame is hung vertically and can hold up to 250 large scallops. This tends to keep the scallops apart. Book nets are made of two meshed wire frames possibly 15 x 50 cm and hinged at one end (Fig. 95). The scallops are laid on one "page" of the book, and the other one folded over, fastened and hung vertically. The pocket and book type culture prevent movement of the scallops, inhibiting shell edged breaks and biting.

Still another method of scallop culture needing minimum equipment is the ear hanging procedure. The scallop's left anterior ear (wing, auricle) is pierced with a 1- or 2–mm hole and attached to a vertical rope with wires or with twine (Fig. 96). The shape of the scallop grown in this way is said to be more convex, with higher meat production and high survival. The method is labour intensive and not suitable for areas with exposure to excessive wave action. Scallops 3 to 5 cm long may be attached to the rope with wire loops, or sewn singly or severally to the rope with monofilament nylon line. The ropes need to be no more than 6 to 8 mm in diameter and any suitable length, as long as they do not touch the bottom. The scallops are placed on the rope at intervals of 10 to 20 cm depending on their size.

Fouling

In bottom culture, normally only the upper left valve is exposed to fouling, for it forms an excellent substrate with deep ribs. It receives all the fouling organisms associated with oysters, such as barnacles, mussels, ascidians, sponges, hydroids, tube worms and some seaweeds. Some species, such as mussels, provide competition for food, but most are nuisance organisms, making shucking more difficult. Included as fouling organisms are borers such as the sponge, *Cliona*, and the polychete, *Polydora*. Effects are similar to those described for oysters.

If fouling becomes excessive, it may interfere with the scallop's mobility. It is impossible to institute any form of fouling control for bottom culture of scallops. Control is possible, but costly, for suspended culture. Fouling becomes particularly important on spat collectors and growing trays, cages

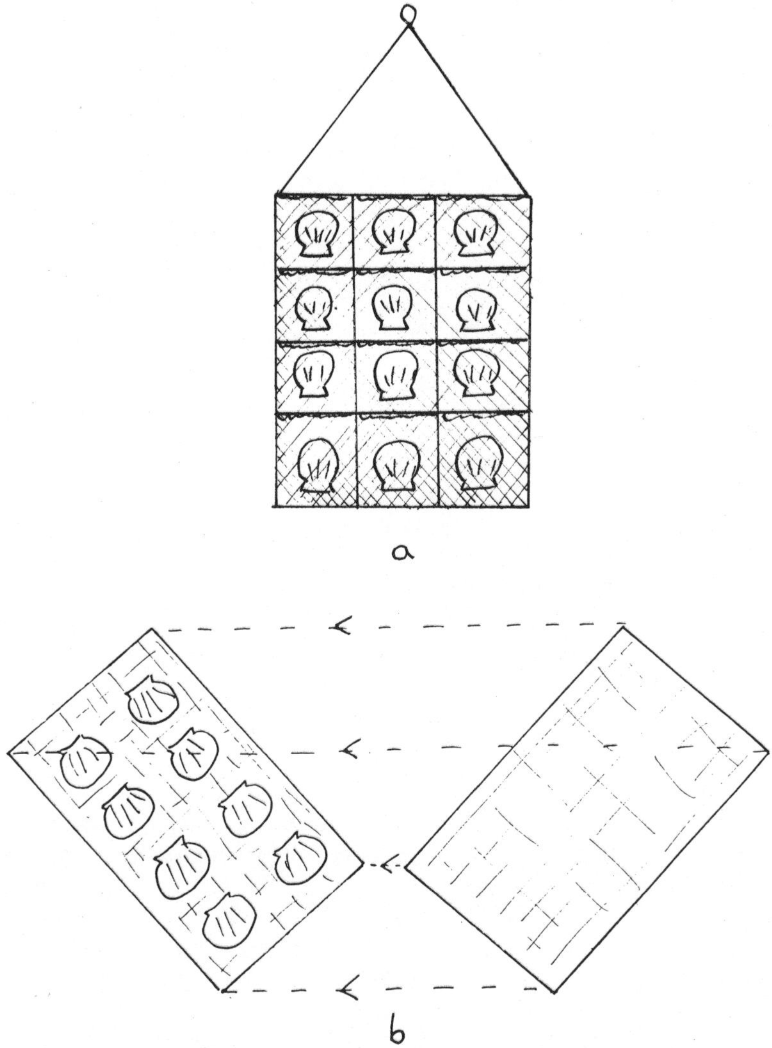

Fig. 95. a) Pocket net and b) book net for culturing scallops.

or strings. Some control may be exercised by adjusting the depth of culture. This is more easily done with suspended culture.

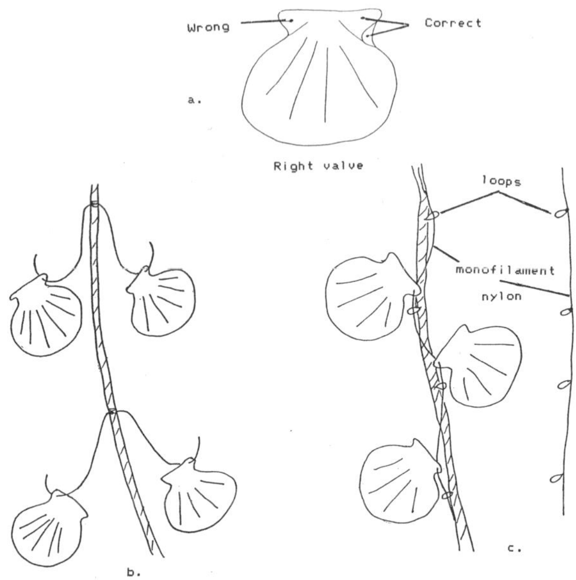

Fig. 96. Ear hanging method for culturing scallops. a) The correct position for the hole is in the anterior ear. b) Wire hooks attached to ropes. c) Monofilament nylon line strung through the ears and the loops of a second monofilament line which has been incorporated into the rope.

PARASITES

Scallops, compared to most bivalves, are relatively free of parasites and those that do occur are not universal to all species. There is a sporozoan, *Pseudoklossia*, of the family Aggregatidae that occurs in one species of scallop. There are a few nematodes such as *Paramisakis* but most are juveniles or larvae. The free living nemertian (*Cerebratulus*), known to attack clams, has also been associated with scallops in Peru on soft bottoms. Pinnotherid pea crabs so common in some bivalves are also found in scallops, but do not cause significant damage except possibly to the gills. Another crustacean that occurs in a Japanese scallop is a species of *Sacculina*, a parasitic barnacle, normally found on crustaceans. This parasite develops a root-like system which, in scallops, occurs near the adductor muscle and is deleterious enough to reduce the value of the host. The gastropod *Odostomia*, mentioned elsewhere, is also parasitic on some scallop species.

CHAPTER 10

TOWARDS CLEANER SHELLFISH: PUBLIC HEALTH ASPECTS OF BIVALVE CULTURE

Safe utilization of molluscan bivalve shellfish is affected by sewage pollution, sanitary handling and a phenomenon known as paralytic shellfish poison or PSP. The solution to the problem of sewage pollution is relatively simple in principle for it is directly related to the bacterial content of the water in which the shellfish are grown. The problem with PSP is that there is little known about the factors affecting the causative agent and thus it is less likely to be controlable. There should be constant surveillance by appropriate public health authorities, both of the waters and of the shellfish entering the market, to insure adequate sanitation of the product to the consumer. In practice, the only immediate solution to both problems is caution. The bivalve farmer should be aware of the consequences of both sewage pollution and PSP and be careful about his product.

PARALYTIC SHELLFISH POISON

Paralytic shellfish poison, generally referred to as PSP, is a difficult problem. Its prevention is impossible and forecasting its occurrence is difficult.

Symptoms

Mild symptoms of PSP, after eating toxic shellfish, take the form of tingling or numbness in the lips and tongue followed by similar sensations in finger tips and toes, usually in 1 h after consumption. More severe symptoms are incoherent speech, prickly feeling in arms and legs with lack of co-ordination in the limbs. Extreme symptoms are muscular paralysis, breathing difficulty and shaking sensations. Fatality is caused by respiratory failure and heart problems. Death times may vary between 3 and 12 h after poisoning.

So far there is no known antidote. Treatment consists of emptying the stomach with an emetic or induced vomiting and a rapid laxative. Artificial respiration should be applied if breathing becomes difficult. Treatment may be given as for primary shock. No alcohol should be given because it aids in the body's absorption of the poison. Warm water in quantity assists in the elimination of the poison in urine.

There are 11 symptoms of PSP. In order of appearance, they are as follows:

- Numbness of the face and lips
- Difficulty with speech
- Numbness of fingers and toes
- Dizziness or giddiness
- Numbness of arms and legs
- Difficulty with breathing
- Difficulty with standing or sitting
- Stomach pain
- Vomiting
- Headache
- Backache.

Domestic animals such as dogs and cats and birds can die from eating toxic shellfish, while cold–blooded vertebrates such as fishes and amphibians are generally insensitive.

Source

The source of poison is a microscopic, one–celled plankton organism belonging to the order Dinoflagellata, which are motile, coccoid or filamentous. When oceanographic and weather conditions occur in the appropriate combination, these organisms can multiply rapidly and become so abundant the sea becomes coloured, usually red, brown or green but other colours also occur.

The species may persist, either as a spore which may lie dormant on the bottom or in the plankton throughout the year, but it is only when the right combination of light, temperature, salinity, nutrients, minerals and vitamins occurs that a bloom develops. In a broad sense these conditions may occur fairly frequently but the specifically correct combination only rarely.

The toxin in dinoflagellates is transferred by extraction and storage to filter feeding animals, mainly molluscs, after ingestion. A filter feeder, depending on size, may become somewhat toxic when the concentration of the toxic dinoflagelate approaches 200/ml in the plankton.

The poison itself, one of the most potent known, affects the transmission of nerve impulses (a neurotoxin); 3000 cells of a *Protogonyaulax* weighing 100 mg wet can yield 15 mg of dry extract or 1 mg of pure poison. The abundance of this species may reach 50,000 organisms/ml, but colouration appears at about 20,000 cells/ml. Such developments are called blooms, and small scale occurrences can develop in less than 1 h and may disappear almost as rapidly. Because of colouration due to the xanthophyll peridinin, such blooms are often called red tides. There are a number of dinoflagellates that may cause PSP, most of which belong to the genus *Protogonyaulax* (formerly *Gonyaulax*).

Many blooms, depending on species, are innocuous, while others may cause invertebrate or fish kills or respiratory irritation in humans. PSP blooms occur throughout the world, but the northeast and northwest coasts of North America are areas of persistent and high level occurrence. In the tropics, areas of concern are Papua New Guinea, Malaysia (Sabah) and Venezuela. The toxin is known variously as *Saxitoxin* (after the butter clam *Saxidomus* on which the early studies on PSP were based) or mytilitoxin (from the mussel which is the source of many PSP difficulties in humans). The toxin, originally considered as a single entity, is now known to be a complex of poisons. Fish kills related to plankton blooms may result from oxygen depletion resulting from decay of dying organisms rather than the poison.

Anatomical Distribution of PSP Toxin in Molluscs

Anatomical distribution of the toxin varies between species and in time after initial toxification. In most species, the digestive gland, gills and mantles are the sites with the high values. In some species, such as *Saxidomus giganteus* of the northwest coast of North America, the siphons contain a considerable portion of the toxin at the time of toxification and can retain it for up to two years.

During initial stages of toxification, the body of the *Saxidimus* may contain 60% of the toxin in the whole clam, but in 6 weeks there can be a complete reversal with the body containing only 15% of toxin with the high proportion in the siphon–gill complex. Gastropods are not filter feeders but may become poisonous by feeding on toxic bivalves.

Measuring PSP

A difficulty in dealing with PSP is the quantitative measurement of the toxin. After much effort and many years of research there is still no rapid way of doing this. The standard procedure, first developed in 1935, is to mince the meats of a sample of bivalves (possibly a dozen, depending on size) and extract the toxin from a known quantity of meat (usually 100 g) by boiling it in weak hydrochloric acid. The volume and acidity are adjusted and the liquid clarified, after which a small quantity is injected peritoneally into the body cavity of a mouse. The average lethal dose, termed a "mouse unit", equals one when 1 ml of extract kills a 20 g mouse in 15 min. If there are 200 ml of extract, then the toxicity would be 200 mouse units/100 g of meat.

By using a purified poison extract as a standard reference, it was possible to express the toxicity in terms of weight (µg) rather than mouse units. The equivalence between mouse units and weight depends on the strain of mice used in the bioassay, but generally about 0.20 µg equals one mouse unit. The accuracy of the test is estimated to be ± 20% and low toxicity values (80 µg/ 100 g) may be underestimated by as much as 60%. The amount of toxin in 3000 *Protogonyaulax* cells is equivalent to one mouse unit. Since an averaged size oyster can filter up to 20 litres of water/hr, and with the possible concentrations of *Protogonyaulax* in the plankton (up to 50,000/ml), it is possible for an oyster to ingest a considerable quantity of toxin. Details of this bioassay may be found on page 319 of Association of Official Analytical Chemists (1975).

Several chemical methods of assay have also been developed but none has so far superceded the mouse bioassay for common use. The method has shortcomings, one being the necessity for a pure strain mouse colony and this is a project in itself.

Partly because PSP is often considered a problem of Public Health Departments and because this authority often has mouse colonies for other purposes, the bioassays are conducted by that group. Departments of Fisheries often co-operate in the collection of samples.

Control Measures

So far no entirely safe method of PSP control has been devised beyond bioassay of the filter feeding molluscs before they are marketed. This is costly and time consuming. Sampling plankton, reporting visual "red tide" occurrences, area and seasonal closures, and monitoring standard stations have been used, but each has its limitations. Unless a plankton program is extensive in terms of sampling, frequency and number of stations, there are too many gaps, in addition to the high cost. Visual sightings are useful up to a point, but the missing of crucial occurrences could be disastrous. Closure of areas where PSP occurs with a degree of regularity is effective. This is usually done on a seasonal basis. This method is in use in several countries. The most common control method is the monitoring of a set of standard stations, at least during the time of year when the likelihood of PSP occurrence is high.

Each PSP area has its own peculiarities. In the PSP area on the Atlantic Coast of Canada, the occurrence with regard to area is annually quite regular, with only timing and level of toxicity varying somewhat. Monitoring a single station permits timing of a seasonal closure. On the Pacific Coast of Canada, there is no such regularity in occurrence over much of the coast either in time or place. However, one section of this coast and part of Alaska is permanently closed, owing to the persistence of PSP particularly in one species of bivalve. California and Washington State have annual summer closures in certain areas. Monitoring specific stations has the limitation of dependence on frequency of sampling. High frequency is costly but low frequency permits the possibility of a toxicity occurrence between sampling times.

Processing and Toxicity

Boiling bivalves until the valves open may reduce toxicity by as much as 70%, but the liquid in which the shellfish were boiled contains about 50% of the toxin in the raw meats. Steaming and frying reduce toxicity more than boiling. Cooking for as short a period as 5 minutes reduces toxicity somewhat and longer periods decreases it further to the 70% level. The canning process can reduce toxicity in clams as much as 90%.

Detoxification

After initial toxification, most species, except for *Saxidomus*, lose the toxin in a few weeks. Siphonate forms take a little longer than the non-siphonated forms such as mussels and oysters.

Natural methods include detoxification of live animals *in situ*. This is the normal method, for in most areas repeated blooms are infrequent. Transplanting to non-toxic areas is costly and there may be no assurance they will remain non-toxic. Artificial methods include chemical treatment by adjustment of pH, salinity or temperature, but they are not very effective and are costly. Removal of the more toxic parts such as the digestive gland and siphon-gill complex is a successful and positive method for the siphonate forms, particularly in such species as *Saxidomus*.

PSP studies required for the establishment of control studies include:

1. Identification of the dinoflagellate species responsible for toxicity and its seasonal and areal occurrence and its ecology, i.e., depth, distribution, temperature, salinity, sunshine, sea conditions before and during blooms.

2. The species of bivalves developing PSP, the level of toxicity reached and the time required for identification as well as the anatomical organs such as digestive glands or siphons which might be areas of poison concentration.

3. Acquisition of information of past history of PSP occurrences from

archives, newspaper files and fishermen. Records of PSP outbreaks in the tropics are relatively recent.

4. Establish history of individual cases of PSP such as:
   a. Date, time, duration of illness.
   b. Species of shellfish eaten, how prepared and how much consumed.

5. Symptoms.

6. Toxicity of shellfish eaten.

Public education on the species of concern, the parts of the bivalve body that should be discarded and how cooking affects toxicity should be developed in PSP areas.

SHELLFISH SANITATION

Since bivalves used for food are harvested or cultured generally close to shore they are subject to any materials contained in drainage from land, whether it be from rivers or streams, runoff after rains or direct discharges. By these means bacteria and viruses can contaminate adjacent shellfish beds or culture sites. One population equivalent (fecal sewage production from one individual) is considered to require daily dilution by 225 million litres of pure water to meet required sanitary standards for shellfish.

Bivalves are filter feeders, so they can filter out and concentrate pathogenic organisms just as well as they do normal food organisms. Once ingested by the shellfish, pathogens can persist for long periods of time and bacteria may multiply under the right conditions. These pathogens are also difficult to detect and enumerate in low concentrations.

The reasons for the sanitary control of edible shellfish operations are:

1. Polluted shellfish can cause illness

2. Variable efficiency of sewage disposal and treatment plants

3. Increase in population densities

4. Shellfish can be contaminated during processing and shipping.

Organisms

The two organisms of main concern in shellfish sanitation are bacteria and viruses. Bacteria, some of which are normal constituents of the human digestive tract, can cause acute gastroenteritis if occurring in excessive amounts. The bacteria of concern are of the genera *Vibrio*, *Salmonella* and *Escherichia* (the coliform group). Members of the *Salmonella* group can cause typhoid fever, while coliforms and vibrios may initiate acute gastroenteritis or simple stomach upsets.

Shellfish have also been responsible for the transmission of viral hepatitis of the A type (jaundice) which can cause liver damage. Contaminated shellfish have also been associated with outbreaks of cholera. These organisms, particularly the pathogens, are difficult to enumerate in low concentrations. The coliform groups are used as indicator organisms since they occur in abundance and reflect in a relative way the possible concentrations of pathogens from sewage. Enumeration of the coliform group as a whole, and *Escherischia coli* in particular, is used as an indication of the level of pollution in both growing waters and in the shellfish itself.

Tests

Water samples for bacteriological analysis should be collected in clean sterile bottles with care taken to prevent contamination during and after sampling. They should be stored at or below 10° C and testing is done preferably within 1 h of sampling and not later than 30 h.

Presumptive tests are done in replicate test tubes (usually 5) containing laural sulphate lactose broth or lactose broth innoculated with three concentrations of the sample (usually 0.1, 1, and 10 ml). These tubes are incubated at 35 ± 0.5° C and examined after 24 h and then at 48 h. Formation of gas within these time periods indicates a positive presumptive test for the coliform group of organisms. The test is then confirmed by innoculating fermentation tubes containing brilliant green lactose bile broth and incubating these for 48 ± 3 h at 35 ± 0.5° C. The formation of gas within this period confirms the presence of coliform organisms.

Density of the coliforms is expressed as the Most Probable Number (MPN)/100 ml of sample. This is a statistic based on the number of positive tubes in an array and is taken from a set of tables. To test for fecal coliforms, similar innoculation procedures are used with an EC broth medium but at a higher incubation temperature of 44.5 ± 0.2° C. If necessary, the various species of coliforms may be differentiated by a further series of tests.

Similar techniques are used for testing shellfish meats themselves by masceration and innoculation of the tubes by appropriate quantities of meats. For testing the quality of shellfish at the market level, standard plate counts are used. Here, agar plates are innoculated with samples of mascerated meats and the developed colonies counted and converted into a quantitative measure of coliform count.

In addition to the tube fermentation method of bacteriological analysis, it is also possible to estimate coliform density in sea water by means of a membrane filter method. The water sample is filtered, with partial vaccuum, on a specially manufactured membrane filter with a rated pore diameter for complete retention of coliform bacteria. It may be necessary to sterilize the filters. Absorbent pads are saturated with Endo type media in special petri dishes and the filter membrane placed on the pad. These are incubated at 35 ± 0.5° C for 24 h. Colonies of bacteria are produced and those with a golden green metallic sheen are considered to be the coliform group. The coliforms are estimated from the count of colonies which should be in the region of 20 to 100/filter. There are also membrane filter procedures for fecal coliforms.

Sanitary Control

The sanitary control of bivalve shellfish is based on bacteriological examinations, supported initially by sanitary surveys. The intent and plan of the bacteriological study will depend to a considerable extent on the findings of the sanitary survey. The sanitary survey is a study of the sources of pollution in the area and how the pollution is distributed. It examines such factors as:

1. Topography, geology and oceanography of the area
2. Natural and artificial drainage systems
3. Rainfall and seasonal rates of stream discharge
4. Shipping channels and moorages
5. Population distribution in place and time
6. Industrial, institutional and recreational areas
7. Economic status of the community
8. Sewage treatment and disposal.

With these data, the design of the bacteriological survey may be undertaken and attention given to:

1. Bacteriological quality of discharges such as streams and storm waters
2. Statistical validity of possible sampling techniques
3. Seasonal factors
4. Effects of hydrographic factors
5. Levels of pollution caused by the main sources
6. The distribution of bacteriological levels in the shellfish of the area.

Classification of Growing Waters

Based on the results of these studies, there are four possible classifications of shellfish producing areas.

1. Approved: Shellfish may be harvested and marketed directly without either treatment or purification. The median coliform MPN should not exceed 70/100 ml.

2. Restricted: This is an area with moderate pollution levels where the median coliform MPN should not exceed 700/100 ml. Shellfish from restricted areas may be used for purification under certain specified conditions.
3. Closed: Such areas are grossly polluted where the median coliform level is in excess of 700/100 ml. Shellfish from such areas may be removed only for relaying or if the shellfish are of seed size.
4. Conditionally approved: This situation may occur as the result of a radical environmental change or production of chemicals or the presence of paralytic shellfish poison. Harvesting is permitted only at times when conditions would be considered safe.

Of importance is the frequency of bacteriological and sanitary surveys. An interval of 2 years is considered a minimum unless the monitoring indicates a re-evaluation. An equally significant feature in shellfish sanitation is that of adherence to the control measures. Success is largely a matter of an adequate control program and surveillance. Education of the producers and consumers is important to guard against unscrupulous marketing of unsafe bivalves.

Purification

It is possible to remove sewage contaminants either by natural purification or by so called artificial means. Another name for the purifiction process is depuration.

Natural purification involves moving, called "relaying", contaminated shellfish from a polluted growing area to an uncontaminated area. Most jurisdictions require a relaying time of two weeks, but the actual purification is accomplished within 2 or 3 days.

In the feeding process in clean water, the contents of the digestive system are removed and the mucous on the gills, mantle and body renewed. The main costs of relaying are due to double harvesting and transportation. This can be about 5% of the cost of producing oysters.

Costs may be reduced somewhat by unitizing the relaying procedure with the use of containers such as wide mesh baskets. Studies on relaying with baskets 62 x 62 x 30 cm holding 55 kg of oysters (*Crassostrea gigas*), and easily carried by two people, showed adequate purification within 48 h over a range of temperature. There is a slight increase in purification rates at higher temperature levels. Even grossly polluted oysters may reach equilibrium with the coliform counts in the purifying environment. Shellfish can be transported from the purification to the processing site in the same container.

Artificial purification involves holding the shellfish in tanks of purified water. This is obtained by purifying raw sea water with ozone, chlorine or ultraviolet light, with the qualification that the water to be treated should not be overly polluted. Purification is virtually completed within 48 h at normal temperature and salinity.

With capital costs, operating costs and bacteriological control, artificial purification can be 10 to 15% of the cost of producing shellfish. Artificial purification of shellfish is widely practiced in Europe. All shellfish in Great Britain are purified before being marketed. In Spain, where mussel production has reached more than 50,000 t per year, all of this quantity is artificially purified before being marketed.

Relaying, where possible, is the economic solution to shellfish purification. The objective is to search for and maintain unpolluted growing areas.

After harvesting, shellfish are subject to storage, processing and shipping. Without suitable precautions, the bacteriological quality can rapidly deteriorate to an unacceptable level. Wet storage requires a water quality equal to approved shellfish growing areas. Dry storage requires refrigeration and speedy utilization. Shellfish transported and sold whole in the shell is termed shell stock. Producers using this method should be certified and each shipment should be labelled in case of problems so its origin may be traced.

Otherwise shellfish are processed in a plant where, at least initially, shucking (opening) and packing is carried out. Here again, certification of the plant is required, for cleanliness and health of the personnel are factors in maintaining a satisfactory bacteriological level.

In summary, shellfish sanitation is an important phase of a shellfish industry requiring a level of technological competence which may only be reached by special training. Health problems created by eating unsanitary shellfish can harm or destroy an industry for a long period of time.

APPENDIX I

EQUIPMENT

VESSELS

The boats required for beginning the study of basic biology and alternative culture methods of bivalves are common to most areas, temperate or tropical. Differences occur in the ability of boats to cope with varying distance, weather and sea conditions. The suitability of a vessel will contribute to the success of a culture project. This is usually the most costly item in a project.

Factors to be considered in the choice of a vessel are:

1. Initial cost
2. Maintenance
   a. Simplicity
   b. Operating manpower
   c. Refit costs
   d. Availability and cost of spare parts
   e. Haul-out facilities and cost
   f. Fuel consumption and cost
3. Seaworthiness
4. Stability as a work platform
5. Carrying capacity — labour, material and equipment.

The key to vessel selection is simplicity. Choose the minimum size that will perform the task adequately with minimum attention. A boat may easily become a problem if it requires so much attention that it interferes with the culture operations. Although it is wise to plan ahead, an investment in a large vessel may be premature during the initial stages.

It is difficult to generalize but one of the more useful types of boat for shellfish work is a fibreglass catamaran type hull about 5 m in length with a 15–hp outboard motor. These are shallow draft, stable, with good carrying capacity and quite safe, though uncomfortable at speed in choppy water. Another advantage is that it may be carried about on a trailer. If distances are great with stormy waters, it may be necessary to use a larger vessel to transport the smaller one but, if possible, site selection should make this unnecessary.

A useful piece of equipment is a small planked or plywood scow about 3 m long, 2 m wide and 0.5 m deep with inclined bow and square stern. This may be towed or powered with an outboard motor.

CULTURE

Equipment will vary somewhat with the type of culture being developed and with what is available locally, but the following list includes the basic needs:

1. Rope, various sizes
2. Wire, 12–14 gauge, galvanized
3. Nails, assorted
4. Lumber
5. Anchors
6. Floats, barrels, bamboo, manufactured (i.e., glass balls, scotchmen, tires filled with urethane)
7. Baskets, plastic, wire or rattan
8. Buckets, plastic or galvanized
9. Oyster-shucking knives
10. Rubber gloves.

The major oyster industries of the world are the result of many years of development and have reached the stage where further refinements in efficiency are considered to come about only by sophisticated study techniques. Such refinements, however, are not for a beginning industry where the objective is the adaptation of an established system of culture. Knowledge necessary for this may be obtained with relatively simple equipment, and complicated apparatus will only divert energy from the main task of establishing the basics in biology and culture. Often funds are available for short time periods. The following lists of tools and scientific eqipment may seem meagre, but they are quite adequate for the initiation of a mollusc culture project. The scientific equipment marked with an asterisk (*) is needed for a research–oriented project but is also recommended for a serious commercial development project. A low–budget project may not need this equipment, depending on the studies planned.

CARPENTER TOOLS

1. Saws, crosscut, rip-Swedish
2. Claw-hammer
3. Sledge hammer
4. Axe
5. Shovels
6. Machete
7. Tin snips
8. Wire snips
9. Crow bar
10. Pliers
11. Files
12. Carborundum stone
13. Knife
14. Tape measure.

OPTICAL EQUIPMENT

1. Stereoscopic stage microscope* magnification to x70
2. Compound microscope* magnification to x400
3. Microscope illuminators*
4. Filar micrometer*
5. Micrometer slide*
6. Enlarged Plexiglass stage for stereomicroscope.*

GLASSWARE

1. Watch glasses*
2. Watch glasses, Syracuse
3. Finger bowls
4. Plastic graduate cylinder of 50, 100, 500, 1000 ml.
5. Jars, 4 oz, 8 oz, 16 oz (0.12, 0.24, 0.47 litre)
6. Dropping pipettes
7. Glass slides and coverslips.*

PLANKTON EQUIPMENT

1. Nets*, #20 (76 - µm mesh), #25 (64 - µm mesh) mesh nylon
2. Plankton buckets* (end of net)
3. Cord*
4. Plankton pump*
5. Counting cell.*

OCEANOGRAPHIC EQUIPMENT

1. Hydrometer or, preferably, salinity refractometer
2. Thermometers
3. Thermographs*, preferably submersible (Ryan type)
4. Current drogue.

CHEMICALS

1. Formaldehyde
2. Methyl alcohol
3. Davidson's fixative*
4. Canada balsam*
5. Xylol*
6. Bleach (Perfex, Chlorox).

LAND TRANSPORTATION

1. Vehicle (pickup truck type)
2. Boat trailer.

MISCELLANEOUS

1. Electronic balance*, capacity 3000 g
2. Triple beam balance
3. Suspension type scale, capacity 10 kg
4. Veeder Root Type Counter.

APPENDIX II

MICROSCOPICAL TECHNIQUE

Certain phases of oyster culture and biology require a microscope. There are many types of microscopes and some, such as electron or scanning types, are able to magnify many thousand times. Higher power compound microscopes have limited usefulness in oyster culture projects and as an equipment item have low priority. For most oyster culture operations, magnifications of 100x or less are quite adequate. The stage stereomicroscope is versatile and most satisfactory. This is a binocular microscope, most of which have combinations of eye-pieces and objectives which give magnifications up to 70x. The oculars should be wide-field and to ease eye strain from prolonged use. The ocular magnification should not be greater than 10x. Stereomicroscopes produce a direct image rather than a reversed one as with higher power compound microscopes, and the working distance is fairly long, i.e., the distance between the objective and the object. Thus, the surface of fairly large objects such as an oyster shell may be viewed easily.

Lighting may be transmitted (reflected) or incident (direct). Reflected light is obtained from a light source and reflected with a substage mirror (a condenser in the case of compound microscopes) and used mainly for oyster larvae and transparent or semi-transparent objects. Incident or direct light illuminates the upper surface of an object (such as an oyster shell) and the light source is above. Microscope lamps are adapted to particular microscopes and the same lamp may be used for both transmitted and incident light.

To make it possible to use large counting cells and examine other large objects such as cultch pieces, a transparent plastic plate can be mounted on the stage and attached with counter-sunk stainless steel or brass bolts through the slide clip holes provided on all stereomicroscope stages. (This also helps to protect the base from seawater.) A plate of Perspex 6 mm thick and 30 x 20 cm is a useful size.

To use a microscope effectively, materials to be observed must be properly prepared. Methods for examination of larvae are described in Chapter 2.

Living material, if small enough, may be observed by placing it on a microscope slide with or without a cover glass depending on its size. Well slides, which have a slight depression in them, are most useful. For these materials transmitted light is generally suitable. Varying light intensity assists in delineating internal structures. Larger solid materials such as pieces of cultch or fouling organisms (barnacles, tunicates) are best observed with incident light. Minute living material may be difficult to observe because of movement as a result of ciliary action. Narcotizing with 10% alcohol or a weak magnesium chloride ($MgCl_2$) will slow down the rate of ciliary action. Both are added gradually.

For fine dissections or probes, dental equipment is useful. A dentist will be glad to donate discarded instruments such as probes and fine needles. Fine scalpels may be purchased but they may also be made by breaking up safety razor blades and mounting the appropriately shaped pieces in a cleft stick or pen holder.

FIXATION AND PRESERVATION

For certain oyster studies, such as the state of reproductive organs or details of internal parasites, it may be necessary to prepare very thin slices of oyster tissue on microscope slides for examination with a compound microscope. To cut thin slices, called sections, the piece of oyster tissue is placed in a block of wax. Wax, however, does not mix with water so it is first necessary to replace the water in the tissue with a substance, called a clearing agent, that will mix with wax. The first stage in this process is called fixation. After fixation and subsequent preservation, material may be sent to a laboratory equipped to do the dehydration, imbedding and section cutting. The complete procedure requires equipment not available except in large laboratories. The small laboratories or projects can do the initial sample preparation. Thus, it is necessary to know something of fixation and preservation. However, similar type studies may be done locally on small objects such as oyster larvae and spat without cutting sections, and these are termed whole mounts.

Fixation is the application of a chemical (fixative) to kill an organism or a part of it and to act on the protein in it so the cellular contents and morphological characteristics retain, as closely as possible, the form possessed in life. Preservation is the maintenance of the fixed condition for extended periods of time. Chemical solutions for fixation and for preservation may be the same, or a different one may be used for the latter.

Fixatives most commonly used for molluscs are Davidson's solution, formol–alcohol, formaldehyde and alcohol. Davidson's is generally used if histological sections are to be made of material such as the gonad for seasonal change studies. Davidson comes in two forms, with acetic acid for fixation of tissue and the other without acetic acid for preservation.

Davidson's with acetic acid

Formaldehyde (40%)	20 parts
Glycerin	10 parts
Alcohol (95%)	30 parts
Glacial acetic acid	10 parts
Water (seawater)	0 parts

After fixation for about 24 h, material may be stored in the same solution but with the deletion of acetic acid.

Formol–alcohol

A satisfactory general purpose combined fixative and preservative with readily available components is formaldehyde–alcohol. The formula is:

Formaldehyde (40%)	100 ml
Alcohol (95%)	900 ml

Another similar, somewhat better fixative is FAA, whose components are:

Ethyl alcohol (50%)	250 ml
Glacial acetic acid	5 ml
Formaldehyde (40%)	3 ml

Formaldehyde

The most widely used fixative and preservative is formaldehyde, also known as formal, formol and formalin. The commercial grade of formaldehyde contains about a 40% solution in water. Typical strengths of formaldehyde used for most molluscan work are 1, 2 and 4% and may be made up as follows, usually with seawater.

	1%	2%	4%
Formaldehyde (40%)	2.5	5	10
Seawater	97.5	95	90

Formaldehyde adequate for most purposes is a 2% solution, providing the volume of formaldehyde is about nine times the volume of the specimen or specimens.

Formaldehyde is acidic and will corrode calcareous structures such as molluscan shells unless it is neutralized or buffered. This may be done by adding borax in excess. Calcium carbonate may be used instead of borax. An alternative is to buffer 20 litre of 10% formalin with 80 g of NaH_2PO_4. Some care must be taken in the use of formaldehyde for the fumes are irritating to the eyes and nasal passages. Some individuals are allergic to this substance.

Alcohol

For fixation and preservation, the type of alcohol generally used is methylated spirits which is less costly than pure ethanol. Isopropyl alcohol is sometimes used, but should be avoided as preservation is poor when it is diluted too much. Alcohol is inflammable and consequently a fire hazard. It also has a high rate of evaporation when used in an open dish for specimen examination. For storage, containers must have tight lids to prevent evaporation and drying of specimens. Industrial methylated spirits are usually purchased most cheaply as a 96% solution.

Dilutions may be made up as follows:

	25% Solution	60% Solution
96% alcohol (ml)	25	60
Water (ml)	71	36
Total volume (ml)	96	96

Thus, the volume of the final dilution in millilitres is the same as the percentage strength of the original alcohol and the amount of alcohol is the same as the final percentage required. This method may be applied to the dilution of any liquid.

For fixing and preserving plankton, a useful combination is 0.5 ml of propylene phenoxetol, 4.5 ml propylene glycol, and 5 ml of 40% formaldehyde in 90 ml of seawater. After fixation, most molluscs or parts of molluscs remain opaque. It is possible to apply chemicals (clearing agents) that will cause them to become transparent so internal structures may be observed. To further delineate these structures, selective staining is possible.

STAINING

Stains for biological materials are numerous and their application may be quite complex. However, for basic molluscan work, methylene blue is a useful stain. A 2 – 5% solution in water is adequate. A 1% solution of neutral red in seawater is used for staining larvae, both alive and fixed.

CLEARING

Clearing agents, generally oil-like, are not miscible with water. Therefore, before an oyster or a piece of tissue may be cleared, the water in it must be removed and this is done by gradually replacing it with alcohol. The usual dehydration schedule, with the time adjusted according to the size of the object is as follows:

1. water	1 – 12 h
2. 50% alcohol	1 – 12 h
3. 70% alcohol	1 – 12 h
4. 90% alcohol	1 – 12 h
5. 95% alcohol	1 – 12 h
6. absolute	2 – 12 h
(100% alcohol — 2 changes)	

Dehydrated objects may then be transferred to a clearing agent such as xylene, clove oil or benzene, all of which are miscible with absolute alcohol. The material is kept in xylol until it is cleared; for objects less than 5 mm, about 1 h may be required. Since xylol causes brittleness, it should not be used for storage but clove oil is satisfactory.

The object may be examined microscopically in this condition in an open dish. However, if a more permanent preparation is required, the object may be mounted on a microscope slide in what is called a mounting medium such as Euparal or Canada Balsam, the latter being in general use and readily available. Clearing agents such as xylol and clove oil are miscible with the mounting materials so the object may be transferred directly from xylol to Canada Balsam. If the object is fairly large a glass ring may be first glued to the glass slide. The ring is filled with Canada Balsam and the object inserted in it and then sealed with a cover slip, taking care not to entrap air bubbles in the process. If the object is fairly flat, narrow strips of glass may be glued to the microscope slide and the object placed between them and covered with Balsam followed by a cover slip.

As an example, to make permanent preparations of oyster larvae, after being concentrated and removed from the plankton sample, they are first narcotized with slowly added crystals of magnesium chloride. This should require only a few minutes, after which they are fixed in 10% formaldehyde in sea water for 1 h. The larvae are then dehydrated by holding them for 10 minutes or so in successive solutions of methylated spirits of 50, 70, 90, 95 and 100% concentration. They may then be mounted directly on a microscope slide, preferably a well type, in Canada Balsam after clearing in xylene. Preliminary staining with neutral red (in water before dehydrating) or Orange G in 95% alcohol may assist in differentiating various larval structures.

For examination of larvae in various positions, they may be mounted in glycerine jelly. The larvae may be placed in the jelly directly from water and may be oriented with a warm needle as required. Glycerine jelly is made up with 10 g gelatine, 70 ml glycerol, 0.25 g phenol crystals and 60 ml distilled water to mix. These components are warmed. The jelly sets firm and is re-warmed for application to a slide.

NARCOTIZING AGENTS

To study some anatomical features, it is often necessary to relax molluscs by means of drugs or narcotics. Common narcotics for marine animals are the magnesium salts, either chloride or sulphate. This is accomplished by the gradual addition of crystals or prepared solutions, 7% magnesium chloride or 20% magnesium sulphate in sea water. Menthol crystals which act rather slowly are also useful. Propylene phenoxetol may be added slowly either as single drops (2 or 3 drops/litre), which sink to the bottom and diffuse from there, or as a mixed 1% solution in seawater. Molluscan larvae are narcotized within 24 h.

Propylene phenoxetol is also used, after formaldehyde fixation, as a preservative and bactericide. With propylene glycol (0.5 g of propylene phenoxetal and 4.5 ml of propylene glycol in 95 ml of seawater), a fungicidal preservative that prevents specimens becoming brittle is formed.

MICROSCOPE MEASUREMENTS

There may be occasions when it is required to measure dimensions of oyster larvae. There are special but costly instruments which are available but a simple method is with an ocular or a Filar micrometer.

An ocular micrometer is simply a round glass disc on whose surface a scale is etched. Divisions and lengths of the scale may vary. The micrometer is inserted in the microscope by removing one ocular, unscrewing the knurled ring at the base, inserting the ocular micrometer, right side up so the figures may be read correctly, and then replacing the knurled ring. The object to be measured is placed along the axis of the scale and the number of divisions on the scale are read off or counted.

These divisions on the ocular micrometer are only relative and must be converted to absolute lengths. This is done by relating the number of divisions in the ocular to an actual distance scale placed on the microscope stage. The distance is on a micrometer slide, which is a glass microscope slide with a 2-mm scale etched upon it. The micrometer eye piece scale is lined up with the scale on the micrometer slide and the number of divisions on the ocular is counted to correspond to a specific distance on the micrometer slide. Thus, the 50 ocular divisions may equal 9 divisions on the micrometer scale, which is in actual distance 0.9 mm or 900 μm. Therefore, one ocular division = 900/50 = 18 μm. So, the oyster larva which measured 25 ocular divisions would be 25 x 18 μm = 450 μm. It must be remembered, however, that the eye piece must be calibrated for each ocular and each microscope objective used. It is wise to calibrate all oculars and objectives likely to be used at one time.

Another type of micrometer is the Filar type. It is somewhat more costly than the eyepiece but more accurate and less time-consuming to use. This consists of an eye piece with a built-in scale or grid and a knob calibrated from one to a hundred. Turning the knob causes a hairline, called the cursor, to move back and forth across the scale. The Filar ocular replaces the normal microscope ocular in the microscope. (Thus, when purchasing be sure it is the right size.) To measure the object, a larva is placed along the scale in the micrometer and with the knurled knob the cursor is moved along to the left of the larva. It is noted this lies between 1 and 2 so the reading will be 1 and whatever is read off on the graduated ring on the side. In this case, since the cursor is about half way between 1 and 2, it will read 55 and the figure 155 is recorded. The cursor is then moved to the right with the knurled knob until it lies on the other (right) end of the larva. This lies between 7 and about three–quarters of the distance toward 8. The reading on the graduated ring is 75 so the number recorded is 775. In other words, the distance between the ends of the larva is the difference between the readings, or 775 - 155 = 620 micrometer divisions. Like the micrometer eyepiece, the Filar must also be calibrated against a micrometer slide. In our example it was found that for the objective in use one Filar division was equivalent to 0.72 µm. So the length of the larva was 620 x 0.72 µm = 447 µm.

APPENDIX III

MAKING FERROCEMENT

The coating of floats to prevent corrosion (barrels), shipworm attack (wood) or erosion (styrofoam) is a problem for shellfish culturists using rafts or long lines. For steel barrels a coating of a mixture of pitch and tar is frequently used but this is only a temporary measure. Antifouling paint for wood is costly and protection is of relatively short duration. One possible solution to the problem is ferrocement, which is simply a thin coating of cement reenforced by an iron framework covering the object to be protected. The weight of cement (2.4 t/m^3 or 150 lb/ft^3) reduces the buoyancy of the float but in most cases (e.g., barrels and styrofoam), this is an advantage in keeping the raft low in the water, thus providing more stability than a high floating one would have. Ferrocement has been used for this purpose for many years as well as for boat building.

MORTAR

There are many types of cement. For boat building the correct one (Type 1 - ASTMC150) should be obtained, but for coating a float ordinary construction grade is satisfactory. This type has early high strength. Rapid hardening, such as may occur in the tropics unless precautions are taken, may cause shrinking and cracking of the coat. Of the 150 lb/ft^3 of cement, about 120 lb are made up of the mortar.

Aggregate is the sand or gravel mixed with pure cement and, for ordinary cement work (sidewalks, retaining walls), a ratio of 5 or 6 parts of aggregate to 1 part of cement is used. There are many types of aggregate, but not all are suitable for ferrocement. For this, only fine aggregates passing through a 4.75 mm (3/16 in.) seive are used. These could be natural sand, crushed stone or crushed gravel/sand but with no impurities. Pure sand with no materials such as soil, sticks or leaves is excellent.

The purity of the sand to be used may be tested by filling a small jar to a depth of about 8 cm with the sand. Water is added to a level 2 to 3 cm above the sand and the bottle shaken. It is then allowed to settle for 24 h. The fine material on top of the sand should not exceed 5% of the depth of the sand. Adding a teaspoon of salt will cause settling to be complete in about 4 h.

For ferrocement, the aggregate to cement ratio is 1.5 to 2 parts aggregate (by volume) to 1 part cement. The cement should be kept dry before use and if there are lumps in the corners of the bag it should not be used. Secure fresh cement if at all possible.

WATER

Water quality for ferrocement is important. It should have no clay, loam, acid content, soluble salts, decaying vegetable matter or any other organic material. Seawater should not be used since it makes concrete tend to absorb water. As a last resort, seawater could be used for wetting down during curing. Public water supplies are usually of satisfactory quality.

Chemical admixtures are sometimes used for speeding setting or for strengthening but these are generally not necessary. Calcium chloride or any admixture containing it should not be used as it may corrode the iron reinforcements.

MIX

A high cement content in the mix generally makes a strong impermeable mortar. The cement/aggregate ratio is about 0.56 as stated above and measured either by weight or volume. Above a cement/sand ratio of 0.66, which is one part cement to 1.5 parts of sand, there will be a drop in strength of the mortar.

Mortar mix proportions (based on whole and half bags of cement) are presented in the following table. The weight is given in kg first with lb in parentheses. Sand volume is given as litres first with cubic feet in parentheses. Water is given as litres with gallons + pints in parentheses.

Cement kg (lb) kg (lb)	Dry sand litres (ft^3)	Water litres (gal+pts)	
1/2 bag 21 (47)	32–43 (70–94)	21–28 (0.75–1.0)	7.5 (1 + 5)
25 (56)	38–51 (84–112)	28–35 (1.0–1.25)	9 (2 + 0)
1 bag 43 (94)	64–85 (140–188)	42–56 (1.5–2.0	15 (3 + 2)
51 (112)	76–102 (168–224)	50–71 (1.75–2.5)	18 (3 + 7)
1.5 bags 64 (141)	101–128 (222–282)	64–85 (2.25–3.0)	23 (5 + 0)
76 (168)	114–152 (252–336)	78–99 (2.75–3.5)	27 (5 + 7)
2 bags 85 (188)	128–171 (282–376)	85–113 (3.0–4.0)	30 (6 + 5)
102 (224)	152–203 (336–448)	99–136 (3.5–4.75)	36 (7 + 7)

The water to cement ratio should be kept as low as possible. Water has a considerable effect on the strength of mortar which decreases as the water content increases. Water is needed to set off the chemical reaction of the cement and bring about hydration. Also it lubricates to facilitate the mixing process. In ferrocement, the water/cement ratio is regulated to about 35 or 40 litres of water/100 kg cement (3.5 or 4 gal water/100 lb of cement). Although weight is the most accurate way of preparing cement and aggregates, volume measurements are satisfactory.

In determining the amount of water, the moisture content of the sand must be considered for it can be 4 to 8% if it has been rained on. A wet sand increases bulk by swelling. One cubic foot of partially wet sand may contain only three-quarters of the volume of dry sand.

The water content of the mix should be kept as low as possible, just wet enough to make it possible to work it into the gaps of the mesh.

CURING

Curing ensures that the the process of hydration continues at as constant a rate as possible. Too rapid drying causes shrinkage with cracks, fractures and loss of strength. The ideal temperature for curing is between 10 and 32° C (50 and 90° F). Low temperature reduces the curing rate. Good curing involves keeping the mortar continuously moist for a minuimum of 7 days and preferrably for 28 days if the object being ferrocemented is large. Fine sprays or moist sacking is used to keep the cement moist.

MESH

Thin layers of unreinforced cement tend to shrink and crack but these stresses can be eliminated by an evenly distributed mesh fabric. Such a fabric is chicken wire, with a mesh of 1.25 cm (0.5 in.) of 18–22 gauge wire. More costly welded mesh or metal plasterer's lath may also be used. For boats, there should be 2–3 lb of reinforcement/ft^2 for 1 in. of thickness (0.9–1.4 kg/ 0.09 $m^2$for each 0.4 cm of thickness). To make up this weight of mesh, eight layers of wire should be sufficient for a layer 2.5 cm (1 in.) thick or less.

CEMENTING

The well–mixed cement is laid on by hand or with a trowel and forced in through the wire meshes to make contact with the surface of the object being covered. After 30 minutes or less of drying, the surface may be smoothed out and curing can commerce.

BIBLIOGRAPHY

Alagarswami, K. 1980. Review on production of mussel seed. In: E. G. Silas, (Ed.), Coastal Aquaculture: Mussel Farming, Progress and Prospects. Cochin, India, Central Marine Fisheries Research Institute, Bulletin 29:22-26.

American Public Health Association. 1981. Standard Methods for the Examination of Water and Waste Water. 15th Edition. American Public Health Association, Washington, DC, USA.

Anderson, G. J., M. B. Miller and K. K. Chew. A Guide to Manila Clam Culture in Puget Sound. Washington Sea Grant Program, College of Ocean and Fishery Science, University of Washington, Seattle, WA, USA. 45 pp.

Angell, C. L. 1984. Culturing the spiny oyster, *Saccostrea echinata*, in Ambon, Indonesia. Journal of the World Mariculture Society 15:433-441.

Angell, C. L. 1985. The Biology and Culture of Tropical Oysters. ICLARM Studies and Reviews 13, 42 pp.

Arakawa, K. 1980. Prevention and Removal of Fouling on Cultured Oysters: A Handbook for Growers. (Translated by R. B. Gillmor.) Maine Sea Grant Program, University of Maine, Orono, ME, USA. Technical Report No. 56. 38 pp.

Askew, C. G. 1978. A generalized growth and mortality model for assessing the economics of bivalve culture. Aquaculture 14:91-104.

Association of Official Analytical Chemists. 1975. Paralytic shellfish poison and biological method. In: Official Methods of Analysis, 12th Edition. Association of Analytical Chemists, Washington, DC, USA. 1094 pp.

Bardach, J. E., J. H. Ryther, and W. O. McLarney. 1972. Aquaculture - the Farming and Husbandry of Freshwater and Marine Organisms. Wiley Interscience, NY, USA.

Bayne, G. L. (Ed.) 1976. Marine Mussels: their Ecology and Physiology. Cambridge University Press, Cambridge, UK. 606 pp.

Blanco, G. J. 1956. The stake (patusok) method of oyster farming in the Dagatdagatan Lagoon, Rizal Province. Philippines Journal of Fisheries 4 (1):21-30.

Blanco, G. J., D. K. Villaluz, and H. R. Montalban. 1951. The cultivation and biology of oysters at Bacoor Bay, Luzon. Philippines Journal of Fisheries 1 (1):33-53.

Braley, R. D. 1984. Mariculture potential of introduced oysters *Saccostrea cucullata tuberculata* and *Crassostrea echinata* and a histological study of reproduction of *C. echniata.* Australian Journal of Marine and Freshwater Research 35:129-141.

Breisch, L. L. and V. S. Kennedy. 1980. A selected bibliography of world-wide oyster literature. University of Maryland, College Park, MD, USA. Sea Grant Publication No. UM-SG-TS-80-11. 309 pp.

Broom, M. J. 1980. The management of *Anadara granosa* (L.) as a natural resource. Resource Management and Optimization 2 (1):1-25.

Broom, M. J. 1982. Structure and seasonality in a Malaysian mudflat community. Estuarine, Coastal and Shelf Science 15:135-150.

Broom, M. J. 1982. Size selection, consumption rates and growth of the gastropods *Natica maculosa* (L.) and *Thaia carnifera* preying on *Anadara granosa* (L.). Journal of Experimental Marine Biology and Ecology 56:215-233.

Broom, M. J. 1982. Analysis of the growth of *Anadara granosa* (Bivalvia: Arcidae) in natural, artificially seeded and experimental populations. Marine Ecology, Progress Series 9(1):69-79.

Broom, M. J. 1983. Mortality and production in natural, artificially seeded and experimental populations of *Anadara granosa*. Oecologia (Berlin) 58:389-397.

Broom, M. J. 1983. Gonad development and spawning in *Anadara granosa*. Aquaculture 30:211-220.

Broom, M. J. 1985. The Biology and Culture of Marine Bivalve Molluscs of the Genus *Anadara*. ICLARM Studies and Reviews, 12. 37 pp.

Cahn, A. R. 1950. Oyster Culture in Japan. US Fish and Wildlife Service, Fisheries Leaflet 383.

Cahn, A. R. 1951. Clam Culture in Japan. US Fish and Wildlife Service, Fisheries Leaflet 399.

Carreon, J. A. 1969. The malacology of Philippine oysters of the genus *Crassostrea* and a review of their shell characters. Proceedings of the National Shellfisheries Association 59:104-115.

Carriker, M. R. 1955. Critical Review of Biology and Control of Oyster Drills *Urosalpinx* and *Eupleura*. US Department of the Interior, Fish and Wildlife Service, Special Scientific Report - Fisheries, No. 148. 150 pp.

Castagna, M. and J. M. Kraeuter. 1961. Manual for growing the hard clam, *Mercenaria*. Virginia Institute of Marine Science, Gloucester Point, VA, USA. Special Report in Applied Marine Science and Ocean Engineering, No. 249. 110 pp.

Central Marine Fisheries Research Institute. 1980. Coastal Aquaculture: Mussel Farming, India - Progress and Prospects. India Council of Agriculture Research, Cochin, India. Central Marine Fisheries Research Institute Bulletin 29, 56 pp.

Chanley, P. and J. D. Andrews. 1971. Aids for identification of bivalve larvae of Virginia. Malacologia 11:45-119.

Cheng, T. C. 1967. Marine molluscs as hosts for symbiosis (with a review of known parasites of commercially important species). Advances in Marine Biology 15:1-424.

Cheong, L. and F. Y. Chen. 1980. Preliminary studies on raft method of culturing green mussels, *Perna viridis* (L.), in Singapore. Singapore Journal of Primary Industries 8(2):119-133.

Chew, K. K. (Ed.) 1982. Proceedings of the North American Oyster Workshop. World Mariculture Society, Baton Rouge, LA, USA. Special Publication No. 1, 300 pp.

Chin, P. K. and A. L. Lim. 1977. Oyster culture development in Sabah. Sabah Society Journal 6(3):107-115.

Claus, C. 1981. Trends in nursery rearing of bivalve molluscs. In: C. Claus, N. DePauw, and E. Jaspers (Eds.). European Mariculture Society, Bredene, Belgium. Special Publication No. 7:1-33.

Clayton, W. E. I. and T. T. Pobran. 1981. A Pacific oyster stake culture pilot project in British Columbia. British Columbia Marine Resource Branch, Fisheries Development Report No. 33, 33 pp.

Curtin, L. 1971. Oyster farming in New Zealand. New Zealand Marine Department, Wellington, NZ. Fisheries Technical Report 72, 99 pp.

Davy, F. B. and M. Graham. (Eds.) 1982. Bivalve culture in Asia and the Pacific. International Development Research Centre, Ottawa, Canada. IDRC-200e, 90 pp.

Devlin, I. (Ed.) 1973. Operation Report: Oyster depuration plant - Lady–smith B.C. Report. Industrial Development Branch, Fisheries and Marine Services, Environment Canada and Commercial Fisheries Branch, Department of Recreation and Conservation Victoria, B.C., Canada. 103 pp.

Durve, V. S. 1965. On the seasonal gonadal changes and spawning in the adult oyster *Crassostrea gryphoides* (Scholtheim). Journal of the Marine Biological Association of India 7(2):328-344.

El Naiem, A. G. 1984. Variability in growth of the mother of pearl oyster (*Pinctada margaritifera*) in the Red Sea (Sudan). Master's Thesis, Dalhousie University, Halifax, Canada. 120 pp.

Field, I. A. 1911. The food value of sea mussels. Bulletin of the United States Bureau of Fisheries 29:85-128.

Fitt, W. K., D. R. Fisher, and R. K. Trench. 1984. Larval biology of tridacnid clams. Aquaculture 39:181-195.

Fritz, L. W. 1982. Annulus formation and microstructure of hard clam (*Mercenaria mercenaria*) shells. M. A. Thesis, College of William and Mary in Virginia. 161 pp.

Galtsoff, P. S. 1964. The American oyster, *Crassostrea virginica* (Gmelin). US Fish and Wildlife Service Bulletin 64, 430 pp.

Glude, J. B. 1974. Identification of oysters of the South Pacific Islands. Proceedings of the National Shellfisheries Association 64:11-12.

Gwyther, J. and J. L. Munro. 1981. Spawning induction and rearing of larvae of tridacnid clams (Bivalvia: Tridacnidae). Aquaculture 24:197-217.

Heslinga, G. A., F. E. Perron and O. Orak. 1984. Mass culture of the giant clams (Tridacnidae) in Palau. Aquaculture 39:197-215.

Hidu, H., C. Conary, and S. R. Chapman. 1981. Suspended culture of oyster: biological fouling control. Aquaculture 22(1-2):189-192.

Hidu, H., W. G. Valleau, and F. P. Veitch. 1978. Gregarious setting in European and American oysters - response to surface chemistry vs. waterborne pheromones. Proceedings of the National Shellfisheries Association 68:11-16.

Imai, T. (Ed.). 1977. Aquaculture in Shallow Seas: Progress in Shallow Sea Culture. Translated from Japanese. Published for the National Marine Fisheries Service and the National Science Foundation, Washington, DC, USA by Amerind Publishing Co., Pvt., Ltd., New Delhi, India. 615 pp.

Iverson, E. S. 1968. Farming the Edge of the Sea. Garden City Press Ltd., Letchworth, UK. 436 pp.

Jenkins, R. 1979. Mussel Cultivation in the Marlborough Sounds (New Zealand). David Jones Ltd., Wellington, NZ. 75 pp.

Kafuku, T. and H. Ikenoue. (Eds.). 1983. Modern Methods of Aquaculture in Japan. Elsevier Scientific Publishing Co., Amsterdam, Netherlands. 216 pp.

Kamara, A. B. 1982. Preliminary studies to culture mangrove oysters, *Crassostrea tulipa*, in Sierra Leone. Aquaculture 27:285-294.

Korringa, P. 1976a. Farming the Cupped Oyster of the Genus *Crassostrea*. Elsevier Scientific Publishing Co., Amsterdam, Netherlands. 224 pp.

Korringa, P. 1976b. Farming the Flat Oyster of the Genus *Ostrea*. Elsevier Scientific Publishing Co., Amsterdam, Netherlands. 238 pp.

Korringa, P. 1976c. Farming Marine Organisms Low in the Food Chain. Elsevier Scientific Publishing Co., Amsterdam, Netherlands. 264 pp.

Korringa, P. 1979. Economic aspects of mussel farming. pp 371- 378. In: T. V. R. Pillay and W. A. Dill (Eds.), Advances in Aquaculture. Fishing News (Books), Ltd., London, UK.

Kraeuter, J. N. and M. Castagna. 1980. Effects of large predators on the field culture of the hard clam, *Mercenaria mercenaria*. US Fish and Wildlife Service, Fisheries Bulletin 78(2):538-541.

La Barberra, M. 1975. Larval and post larval development of the giant clams

Tridacna maxima and *Tridacna squamosa* (Bivalvia, Tridacnidae). Malacologia 15(1):69-79.

Loosanoff, V. L. and H. C. Davis. 1963. Rearing of bivalve molluscs. Advances in Marine Biology 1:1-136.

Loosanoff, V. L., H. C. Davis and P. E. Chanley. 1966. Dimensions and shapes of larvae of some marine bivalve mollusks. Malacologia 492(2):351-435.

Lopez, M. D. and E. D. Gomez. 1982. Reproductive cycles of the oysters *Crassostrea echinata* and *C. lugubris* in Calatagan, Batangas, Philippines. Kalikasan 11(1);57-73.

Lutz, R. A. (Ed.). 1980. Mussel Harvest and Culture: A North American Perspective. Elsevier Scientific Publishing Co., Amsterdam, Netherlands. 350 pp.

Lutz, R., J. Goodsell, M. Castagna, S. Chapman, C. Newell, H. Hidu, R. Mann, D. Jablonski, V. Kennedy, S. Siddall, R. Goldberg, H. Beattie, C. Falmagne, A. Chestnut and A. Partridge. 1982. Preliminary observations on the usefulness of hinge structures for identification of bivalve larvae. Journal of Shellfish Research 2(1):65-70.

Maclean, J. L. 1979. Indo-pacific red tides. pp. 173-178. In: D. L. Taylor and H. H. Seliger (Eds.), Toxic Dinoflagellate Blooms, Vol. 1. Elsevier Scientific Publishing Co., New York, USA.

Maclean, J. L. and A. W. White. 1985. Toxic dinoflagellate blooms in Asia: a growing concern. pp. 517-520. In: D. M. Anderson, A. W. Whyte and D. G. Baden (Eds.), Toxic Dinoflagellates. Proceedings of the Third International Conference on ToxicDinoflagellates, St. Andrews, New Brunswick, Canada. Elsevier Publishing Co., New York.

Magoon, C. 1982. An Introduction to Shellfish Aquaculture in the Puget Sound Region. Department of Natural Resources, Olympia, WA, USA. 68 pp.

Magoon, C. and R. Vining. 1981. Introduction to Shellfish Aquaculture. Washington Department of Natural Resources, Seattle, WA, USA. 69 pp.

Mann, R. 1979. The effect of temperature on growth, physiology and gametogenesis in the manila clam *Tapes philippinarum.* Journal of Experimental Marine Biology and Ecology 38:122-133.

Mann, R. (Ed.). 1979. Exotic Species in Mariculture. Massachusetts Institute of Technology Press, Cambridge, MA, USA. 363 pp.

Mason, J. 1983. Scallop and Queen Fisheries in the British Isles. Fishing News Books, Ltd., Farnham, Surrey, UK. 144 pp.

Medcof, J. C. 1961. Oyster farming in the Maritimes. Fisheries Research Board of Canada Bulletin 131.

Milne, P. H. 1972. Fish and Shellfish Farming in Coastal Waters. Whitefriars Press Ltd., London, UK. 208 pp.

Morse, D. E., K. K. Chew and R. Mann. (Eds.). 1984. Recent Innovations in Cultivation of Pacific Mollusks. Elsevier Scientific Publishing Co., Amsterdam, Netherlands. 404 pp.

Morton, J. E. 1960. Molluscs: an Introduction to Their Form and Function. Harper Textbooks, New York, NY, USA. 232 pp.

Mottet, M. G. 1979. A review of the fishery biology and culture of scallops. State of Washington, Department of Fisheries, Technical Report No. 39. 100 pp.

Newkirk, G. F. 1980. Review of the genetics and potential for selective breeding of commercially important bivalves. Aquaculture 19:209-228.

Ng., F. O. 1979. Experimental culture of the flat oyster (*Ostrea folium*) in Malaysian waters. Malaysian Agricultural Journal 52(2):103-113.

Prakash, A. J., J. C. Medcof, and A. D. Tennant. 1971. Paralytic shellfish poisoning in Eastern Canada. Fisheries Research Board of Canada, Bulletin 171. 877 pp.

Purchon, R. D. 1977. The Biology of the Molluscs, 2nd ed. Pergamon Press, New York, NY, USA.

Quayle, D. B. 1969. Paralytic shellfish poisoning in Brithish Columbia. Fisheries Research Board of Canada, Bulletin 168. 68 pp.

Quayle, D. B. 1971. Pacific oyster raft culture in British Columbia. Fisheries Research Board of Canada, Bulletin 178. 34 pp.

Quayle, D. B. 1975. Tropical oyster culture - a selected bibliography. International Development Research Centre, Ottawa, Canada, 052e. 40 pp.

Quayle, D. B. 1980. Tropical oysters: culture and methods. International Development Research Centre, Ottawa, Canada, TS17e. 80 pp.

Quayle, D. B. 1988. Pacific oyster culture in British Columbia. Fisheries Research Board of Canada, Bulletin 218. 241 pp.

Quayle, D. B. and R. Bernard. 1976. Purification of basket-held Pacific oysters in the natural environment. Proceedings of the National Shellfisheries Association 66:69-75.

Quayle, D. B. and N. Bourne. 1972. The clam fisheries of British Columbia. Fisheries Research Board of Canada. Bulletin 179. 70 pp.

Rees, C. B. 1950. The identification and classification of lamellibranch larvae. Hull Bulletin of Marine Ecology 3:77-104.

Rhoads, D. C. and R. A. Lutz (Eds.). 1980. Skeletal Growth of Aquatic Organisms. Plenum Press, New York, NY, USA. 750 pp.

Russell, F. E. 1984. Marine toxins and venomous and poisonous marine plants and animals (invertebrates). Advances in Marine Biology 21:60-217.

Siddal, S. E. 1980. A clarification of the genus Perna (Mytilidae). Bulletin of Marine Science 161(4):858-870.

Sindermann, C. J. 1970. Principal Diseases of Marine Fish and Shellfish. Academic Press, New york, NY, USA. 399 pp.

Sindermann, C. J. 1988. Disease Diagnosis and Control in North American Marine Aquaculture, 2nd ed. (revised). Elsevier Scientific Publishing Co., Amsterdam, Netherlands. 426 pp.

Spencer, B. E. a nd C. J. Gough. 1978. The growth and survival of experimental batches of hatchery-reared spat of *Ostrea edulis* L. and *Crassostrea gigas* Thimberg, using different methods of tray cultivation. Aquaculture 13:293-312.

Squires, H. J., M. Estevez, O. Barona and O. Mora. 1975. Mangrove cockles, *Anadara spp.* (Mollusca:Bivalvia) of the Pacific Coast of Colombia. Veliger 18:57-68.

Stephen, D. 1980. The reproductive biology of the Indian oyster, *Crassostrea madrasensis* (Preston) I. Gametogentic pattern and salinity. Aquaculture 21:139-146.

Thompson, J. M. 1954. The genera of oysters and the Australian species. Australian Journal of Marine and Freshwater Research 5:132-168.

Toral-Barza, L. and E. D. Gomez. 1985. Reproductive cycle of the cockle *Anadara antiquata* L. in Calatagan, Batangas, Philippines. Journal of Coastal Research 1(3):241-245.

Toyo, T., I. Tesuji and N. Inoue. 1978. The mass culture of the ark - Anadara and their problems in Yamaguchi Prefecture. Culture Research 7:51-66.

US Department of Health, Education and Welfare. 1965. National Shellfish Sanitation Program, Manual of Operations; Parts I, II and III. Public Health Service, Division of Environmental Engineering and Food Protection, Shellfish Sanitation Branch, Washington, DC, USA.

Ventilla, R. F. 1982. The scallop industry in Japan. Advances in Marine Biology. 20:310-382.

Ventilla, R. F. 1984. Recent developments in the Japanese oyster culture industry. Advances in Marine Biology 21:57 pp.

Walne, P. R. 1979. Culture of Bivalve Molluscs - 50 Years' Experience at Conwy, 2nd ed. Whitefriars Press Ltd., London, UK. 189 pp.

Walne, P. R. and R. Mann. 1975. Growth and biochemcial composition in *Ostrea edulis* and *Crassostrea gigas*. pp. 587-607. In: H. Barnes (Ed.), Proceedings of the 9th European Marine Biology Symposium, Aberdeen University Press.

Watters, K. W. and T. E. Prinslow. 1975. Culture of the mangrove oyster, *Crassostrea rhizophorae* Guilding, in Puerto Rico. Proceedings of the World Mariculture Society 6:221-236.

Welder, E. 1970. Experimental spat collecting and growing of the oyster *Crassostrea rhizophorae* Guilding in the Cienaga Grande Santa Marta, Colombia. Aquaculture 21:251-259.

Wilkins N. P. and E. M. Gosling. (Eds.). 1983. Genetics in Aquaculture. Elsevier Scientific Publishing Co., Amsterdam, Netherlands. 425 pp.

Wood, L. and W. J. Hargis, Jr. 1971. Transport of bivalve larvae in a tidal estuary. Cambridge University Press, Oxford, UK. European Marine Biology Symposium 4:29-44.

Yonge, C. M. 1960. Oysters. Collins, London, UK. 209 pp.

Yonge, C. M, and T. E. Thompson. 1976. Living Marine Molluscs. Collins, London, UK. 288 pp.

SUBJECT INDEX

TAXONOMIC INDEX